Patrick Moore's  es

For further volumes:
http://www.springer.com/series/3192

Observing the Messier Objects with a Small Telescope

In the Footsteps of a Great Observer

Philip Pugh

Springer

Philip Pugh
Wiltshire, UK

ISBN 978-0-387-85356-7   e-ISBN 978-0-387-85357-4
DOI 10.1007/978-0-387-85357-4
Springer New York Dordrecht Heidelberg London

Library of Congress Control Number: 2011935454

© Springer Science+Business Media, LLC 2012
All rights reserved. This work may not be translated or copied in whole or in part without the written permission of the publisher (Springer Science+Business Media, LLC, 233 Spring Street, New York, NY 10013, USA), except for brief excerpts in connection with reviews or scholarly analysis. Use in connection with any form of information storage and retrieval, electronic adaptation, computer software, or by similar or dissimilar methodology now known or hereafter developed is forbidden.
The use in this publication of trade names, trademarks, service marks, and similar terms, even if they are not identified as such, is not to be taken as an expression of opinion as to whether or not they are subject to proprietary rights.

Springer is part of Springer Science+Business Media (www.springer.com)

This book is dedicated to the great loves in my life: my wife, Helga and daughter, Marcela. They will be in my heart always and forever.

Preface

It is rather strange that Charles Messier's list of objects to avoid when hunting for comets became the first of many definitive lists of deep sky objects to view. He did, indeed, discover eleven comets, yet it is his "catalogue" that is his main claim to fame. Many would argue that the Messier Catalogue has little relevance today, with many amateur astronomers owning telescopes far larger than he had. Indeed, many advanced amateurs prefer to list objects from the New General Catalogue (NGC), which covers fainter objects and those not visible from Europe. I like the Messier Catalogue because I am familiar with many of its objects, some of which can be enjoyed with modest instruments from less than ideal viewing sites.

In fact, my inspiration for the book came from my own ideas for a comet hunt. One night, I was looking around the Lyra/Hercules/Draco region with binoculars, mostly to look at the area's many double stars. I came across a bright object that looked like a comet. Fortunately, I knew better than to go e-mailing the globe with claims of a new discovery. My "comet" was none other than the globular star cluster M92! I already knew of M92 but, as it was an exceptionally clear night and, as I'd never seen it so bright before, I didn't recognize it immediately. Cursing more that I didn't have any suitable imaging equipment, rather than it wasn't a new discovery, I took the opportunity to have a closer look with my 127 mm Maksutov-Cassegrain, known as a "Maksutov" for short.

This book is a personal voyage of discovery. Although many of the "Usual Suspects" (list of deep sky objects that are easily visible to binoculars) are in the Messier catalogue, some I had never seen before I started researching this book, or had seen a fuzzy patch in the place they were known to exist but not much else.

Indeed, there was a time when I even wondered if I would ever complete this book or find myself outside in the freezing winter trying to map the Virgo Galaxy Cluster before dawn, after two successive springs of missing it! Fortunately, the spring of 2009 was clear enough for me to see it. Not only was it necessary to complete this book but was also a personal ambition of mine. My family thought I was mad, though, when I spent several nights awake until 3 a.m. in order to see the elusive members of the Messier Catalogue in southern Scorpius and Sagittarius. In fact, I'd already seen some of them from the southern hemisphere but, as I was trying to follow in the footsteps of Charles Messier himself, that would count as cheating! I finally completed observations of all objects in May 2010.

But first, a foreword by Kulvinder Singh Chadha on the man himself. Thanks also to him for researching Charles Messier's own observing notes.

Wiltshire, UK   Philip Pugh

Acknowledgements

I'd like to thank the following people (in no particular order):

- Kulvinder Singh Chadha for his research into Charles Messier himself. Kulvinder is a named contributor
- Anthony Glover and Mike Deegan for providing images and descriptions. They are also named contributors
- Manchester Astronomical Society for providing images. Individual photographers are named. The society will receive a share of the proceeds of this book
- NASA for providing images
- John Watson, UK editor of Springer for his guidance and advice
- Springer Science+Business Media for being patient with the delays due to the English weather

Contents

1	**Charles Messier: His Life, Discoveries, and Legacy**	1
	Early Life and Education	1
	The Race for Halley's Comet	7
	Clock Watching Beyond the Sea	12
	Recognition, Tragedy and a Big Egg	15
	Revolution and Napoleon	19
	The Legacy of Charles Messier	23
2	**Introduction to the Messier Objects**	25
	The Messier Objects	26
3	**M1–M22**	31
	M1	31
	M2	34
	M3	37
	M4	41
	M5	44
	M6	47
	M7	51
	M8	53
	M9	57
	M10	60
	M11	63
	M12	65
	M13	68
	M14	72
	M15	75

	M16	78
	M17	81
	M18	85
	M19	87
	M20	90
	M21	94
	M22	96
4	**M23–M45**	**101**
	M23	101
	M24	104
	M25	106
	M26	109
	M27	112
	M28	115
	M29	118
	M30	121
	M31	124
	M32	128
	M33	130
	M34	134
	M35	138
	M36	140
	M37	143
	M38	145
	M39	147
	M40	149
	M41	151
	M42	154
	M43	156
	M44	159
	M45	163
5	**M46–M68**	**167**
	M46	167
	M47	170
	M48	172
	M49	175
	M50	178
	M51	180
	M52	184
	M53	186
	M54	189
	M55	192
	M56	195

Contents   xiii

M57	198
M58	201
M59	204
M60	207
M61	209
M62	212
M63	215
M64	218
M65	221
M66	224
M67	226
M68	228

6 M69–M91   233

M69	233
M70	236
M71	239
M72	242
M73	245
M74	248
M75	250
M76	253
M77	256
M78	259
M79	263
M80	266
M81	268
M82	272
M83	275
M84	277
M85	280
M86	283
M87	285
M88	288
M89	291
M90	294
M91	297

7 M92–M110   301

M92	301
M93	304
M94	307
M95	310
M96	312
M97	316

M98	318
M99	321
M100	322
M101	326
M102	329
M103	332
M104	334
M105	336
M106	338
M107	341
M108	343
M109	346
M110	348
Glossary	353
Index	369

Chapter 1

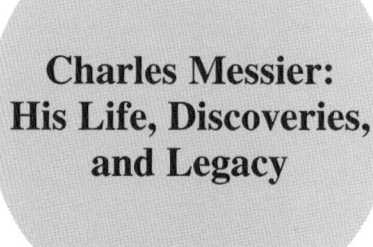

Charles Messier: His Life, Discoveries, and Legacy

Early Life and Education

Charles Messier was born on the 26th of June, 1730, in Badonviller, France, near Strasbourg, close to the German border, one of the youngest of 12 children. He was the tenth child of Nicolas and Françoise Messier, so he wasn't quite the baby of the family.

The Messiers lived in the (then) kingdom of France, near the Duchy of Lorraine. The bloody upheaval of the French Revolution and the rise of Napoleon Bonaparte were still quite a number of decades away.

There is a commonly held view that Charles had an impoverished background, in the material sense. In such circumstances, Nicolas's efforts to provide for such a large family would have been challenging, to say the least. In actual fact it seems that Charles and his siblings grew up in reasonable affluence, though the idea of an impoverished background could be a reasonable assumption to make in the view of such a large family. Debate still continues on whether this was actually the case, but clearly the notion of a successful astronomer growing up in poverty is a romantic one. Nicolas Messier was likely able to provide adequately for his family, indeed even allow them to live in some level of comfort because of his work as an administrator for the regents of Salm, in the court of Salm-Kyrburg (one of many Holy Roman principalities of Germany, France, and Luxembourg). The Regent Philip Joseph and his brother Johan would eventually become the princes of Salm-Kyrburg, which was a German principality that bordered Lorraine on its northeastern side. But in 1741, Nicolas died.

Even as recently in the past as the eighteenth century, death was never far away in everyday life, whether you lived a princely, or a more desperate lifestyle. For the young Charles to have already lost six of his brothers and sisters by the time of his father's death was not unusual for the time. Hyacinthe, the eldest child (and male), now became the head of the family, as was the natural and expected thing to do. But with Nicolas's role in the house of Philip Joseph now no more, how was the Messier family to survive, let alone maintain the level of living that its members were accustomed to?

They need not have worried. The 24-year-old Hyacinthe, like his father, took up an administrative role himself in the princes' household, having returned to Badonviller from the town of Nancy in 1740, where he worked as an assistant to a palace curator. Hyacinthe following his father into the prestigious clerical profession was clearly fortunate for the Messiers. And it was around this time that something fateful happened in the history of astronomy. Not for the only time in his life, Charles, who was then 10 years old, was to have an accident. It is said that while playing in a boisterous manner in the family home he fell out of an open window. It must have been quite a considerable fall, because the young Charles had broken his leg at the thigh. Perhaps it was due to this incident and the inevitable length of time it would take to recover from it that Charles was from then on educated at home, under his elder brother's tutelage. This private schooling was to continue for the best part of a decade, with Charles learning the ins and outs of administrative and clerical work. During that time he developed a keen eye for detail, thanks to Hyacinthe's one-on-one tutoring, which was to prove highly advantageous later on.

It was also during this time that Charles began to demonstrate a nascent interest in astronomy. When he was 14 years old a fantastic comet (discovered by Dutch astronomer Dirk Klinkenberg) appeared in the skies above Lorraine. Designated C/1743 X1, the comet was actually named for the wealthy Swiss mathematician Jean-Philippe Loys de Chéseaux because of his detailed observations of it. It must have been a truly spectacular sight because Comet de Chéseaux was a luminary behemoth, possessing six tails – something that was confirmed by Joseph Nicolas Delisle, France's Astronomer of the Navy. The sight of this celestial gem outshining Jupiter in the night sky probably had a profound impact on the young Charles and quite possibly crystallized his fascination of the heavens (and of comets in particular). Then in 1748, when he was 18, an annular solar eclipse appeared over Badonviller on the 25th of July. An annular eclipse leaves a searing ring of sunlight around the Moon as it passes over the Sun's disk. This meant that Badonviller was not plunged into total darkness, as in the case of a more spectacular total eclipse, but it would still have been a sight to behold. To the young Charles it was as if the heavens were sending him a message as to where his future lay.

And local politics seems to have helped push Charles in that direction, too, albeit in an indirect manner. In 1751 there was a redrawing of political boundaries in the region (much like what occasionally happens with modern English counties) and Badonviller was no longer under the jurisdiction of the House of Salm-Kyrburg.

Early Life and Education

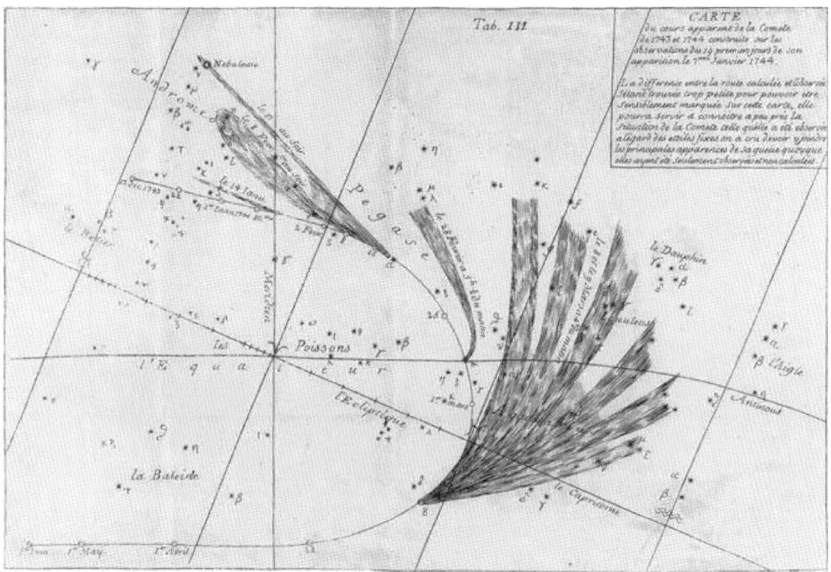

Fig. 1.1 Delisle's depiction of the great comet of 1744 (1 of 3) (Image courtesy of the Paris Observatory)

Being a loyal employee and subject, Hyacinthe left Badonviller again, to follow the Salm-Kyrburgs, and settled in Senones, which today is a small town situated between Nancy and Strasbourg. Charles, now left behind with the remaining members of his family, had to now look for employment. He turned to the trusted Abbot Theolen, who asked around for jobs on Charles's behalf. It was quite common at the time for abbots to act as family mentors and advisers. In fact their role could be considered not too dissimilar from that of modern-day 'agony uncles.' Theolen was clearly effective, finding not one but two potential positions that Charles could have taken up, both of which were in Paris. One was as assistant to a curator, and the other was with the naval astronomer Joseph Delisle, the man who had confirmed Comet de Chéseaux's six-tailed appearance. Charles would have to leave Badonviller for the great city of Paris in either case, but was unsure which of the two job offers he should pursue. Naturally he turned to his older brother for help in making a decision. Did the curator Hyacinthe think that Charles should take on the same role that he and his father had? No. Hyacinthe thought perhaps that Charles should work for Delisle. His wise reasoning for this was that the position would have offered Charles much more in the way of prospects. It was a decision that could not have chimed better with Charles's latent astronomical interests (Figs. 1.1, 1.2 and 1.3).

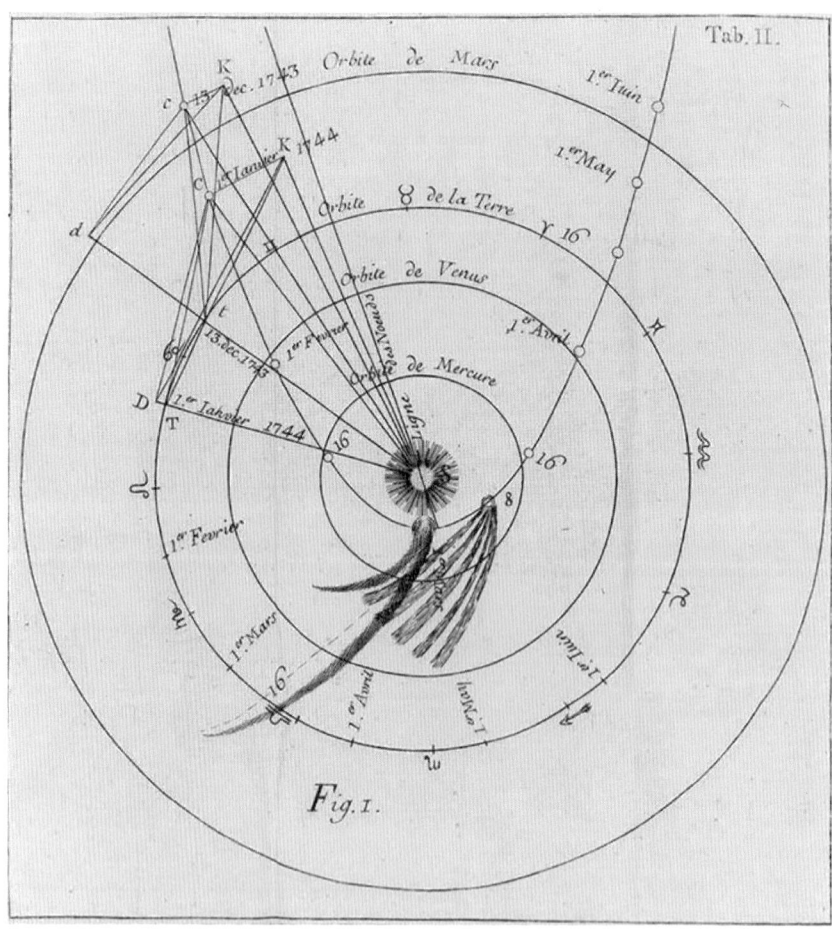

Fig. 1.2 Delisle's depiction of the great comet of 1744 (2 of 3) (Image courtesy of the Paris Observatory)

Charles started work for Delisle on the 2nd of October, 1751, as a clerk-assistant. The building where Delisle was based was called the Hôtel (townhouse) de Cluny; built in 1480 on Roman ruins as a dwelling place for a Parisian order of monks known as the Abbots of Cluny. The building didn't actually belong to the navy, and it was only now in the eighteenth century that it was let to their administration.

When Charles arrived, Delisle was said to have been particularly impressed by his neat handwriting. In fact the astronomer Jean-Baptiste Delambre (known for being a gregarious man) made much of the fact that it was Charles's handwriting

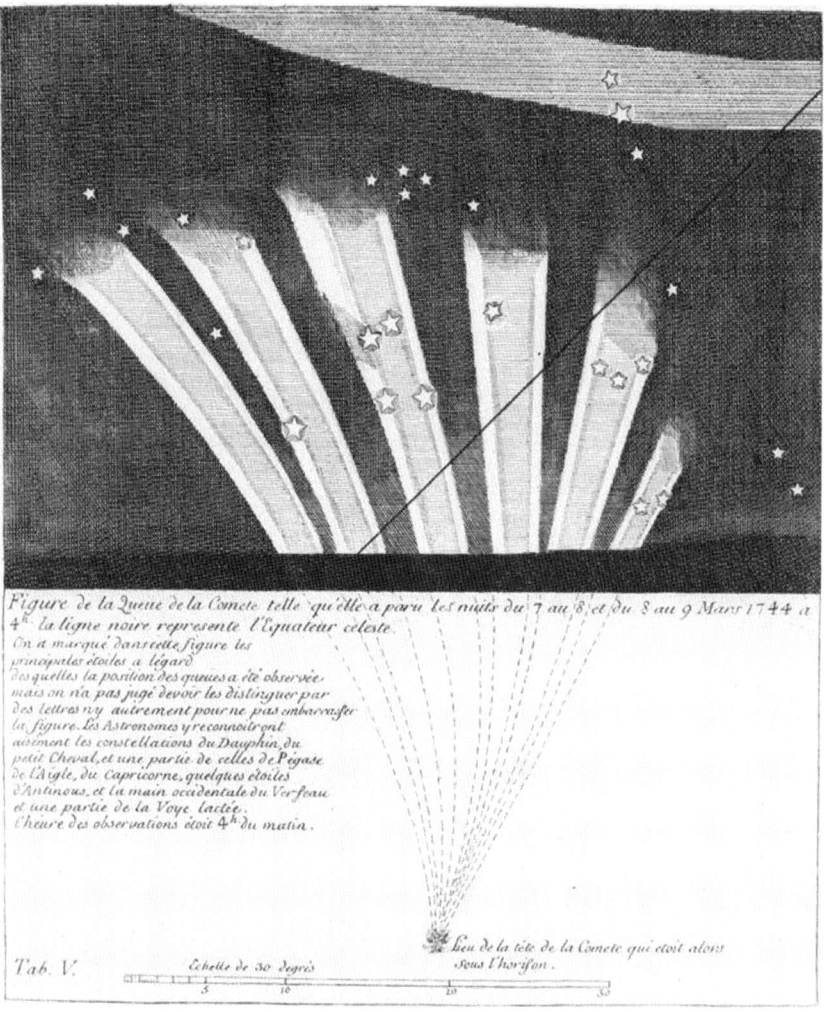

Fig. 1.3 Delisle's depiction of the great comet of 1744 (3 of 3) (Image courtesy of the Paris Observatory)

that secured his position at Cluny. Delambre in later years would write a biography of Charles Messier, and much of what is known about him today comes from this source.

So it seemed that the years of private education by his munificent older brother were beginning to pay off for Charles. Being a childless man in his sixties, Delisle developed a paternalistic bond with the young Charles, who was 21 at the time.

Delisle and his wife offered Charles accommodation in the Royal College of France, where the couple lived. In the space of a few days Charles Messier had made it to Paris, impressed his employer, and now had a place to stay, too. Not bad!

However, when he got down to work Charles's first task wasn't astronomical at all but involved copying out a map of China. Specifically, Charles copied out depictions of the Great Wall, built in 200 BC by the Emperor Huang to protect China and the Old City of Peking – settled since the Iron Age – from Mongolian invaders. Joseph Delisle was a widely traveled man who recently returned from a 21-year stay in Russia where he taught astronomy and even helped to build an observatory in St Petersburg. The old man reputedly had a love of old charts and documents and must have come across the maps of the Great Wall and the Old City during his travels.

As fascinating as this was, Charles wouldn't be kept away from astronomy for long, for he now had a new job: to keep a record of all the observations undertaken at Delisle's observatory, which sat at the top of the Hôtel de Cluny. It was a marvel to see. Built by Delisle with his own hands, it consisted of a wooden structure with glass panels and must have been quite a sight. The Cluny building is now a museum of medieval arts with displays that include many fine tapestries, though tragically, the observatory itself was completely dismantled in the nineteenth century.

Although he was working with astronomical records, Charles wasn't actually doing any astronomy himself, and this is what he really yearned to do. Being in the employee of the Naval Astronomer and working so close to the observatory, how could he not? As it turned out he could not have wished to be in a better place for it and didn't have to wait much longer. Delisle had a personal assistant at Cluny called Libour who showed Charles all the ropes, including how to use the instruments in the observatory. The telescopes housed in the observatory, though large in aperture, were actually not the most efficient instruments around even for their day. This probably didn't matter a great deal to Charles when he finally started observing heavens, for he must in some way have felt overwhelmed with the notion that his time had now come.

The first recorded observation that Charles made at the observatory was the transit of the planet Mercury across the Sun's disc on the 6th of May, 1753. Planetary transits at that time were of great importance in determining the accuracy of orbital calculations, as well as for predicting future transits. It was only since the work of Johannes Kepler in the 1600s that planetary positions could be calculated with any accuracy at all. In his time Delisle was to organize expeditions to various parts of the world for planetary transit observations. These kinds of expeditions were a little-known concept in astronomy in the 1700s, but became much more commonplace in the centuries that followed.

By 1755 Charles was promoted to depot clerk of the navy. Delisle had sold his large collection of books, maps, and documents to the government in order to get an annuity for himself, and Charles didn't miss out either. He got board and lodgings at the observatory for Delisle's efforts, as well as an annual salary of 500 francs (Fig. 1.4).

The Race for Halley's Comet

Fig. 193. — Hôtel de Cluny au xvIII^e siècle. Tour de l'Observatoire de la Marine. (D'après Saint-Victor, Tableau de Paris.)

Fig. 1.4 The grounds of the Hôtel de Cluny, rented to the naval administration of France. The observatory where Charles Messier worked can clearly be seen on top of the column (Image courtesy of the Paris Observatory)

The Race for Halley's Comet

One of the big questions in astronomy at the time was whether comets were random visitors to the Solar System or were they somehow traveling in periodic orbits, much like the planets did. The British astronomer (and later second Astronomer Royal after John Flamsteed) Edmund Halley posed this question in 1701. The Oxford-educated Halley charted sunspots and went on expeditions (much in the same way as Delisle). In this case it was to the southern hemisphere to map the positions of the stars there. But clearly he is best associated with the comet that now bears his name.

Indeed, why would Halley's name be most associated with periodic comets? The reason was simple. When calculating the orbital dynamics of a comet that had appeared in 1681–1682, Halley noticed something curious about his figures. The calculations seemed to match that of a comet seen in 1531 by the German mathematician and astronomer Petrus Apianus, and also to one seen by Johannes Kepler in 1607. The period between these previous two sightings caught Halley's eye too, at 76 years apiece. This was also the same period between Halley's own sighting and that of Kepler's. Coincidence? Halley took an intellectual step and predicted that these sightings were all of the same comet, and that it should return in another 76-odd years – though he knew he wouldn't be around to see it. Up until that time it was commonly accepted that the appearance of comets was a haphazard affair, and now Halley put forward an alternative theory that got comet hunters excited.

So in 1758 Europe's astronomical world was eagerly awaiting the supposed return of the comet predicted by Edmund Halley. It's not clear at the time of writing if other parts of the world had heard of Halley's theory, or indeed discovered it independently. If the comet were to be spotted it would for the first time in history be proof that comets had periodic orbits. And that meant that they had to somehow be part of the Solar System and not chance visitors from beyond.

Halley predicted that the comet would return in late '58 or early '59. Astronomers such as Jérôme Joseph Lalande, who worked at the Hôtel de Cluny, calculated the date of the comet's perihelion as 13th of April, 1759. But this date, like any when calculating positions for comets, was subject to revision. As the exact date of the comet's initial visibility was unknown, astronomers around Europe had started scanning the skies in early 1758. For Charles, this was a great opportunity to make a name for himself, and he started his own search as early as 1757. After all, using Delisle's own orbital calculations he would surely spot it first. So three years into his role as depot clerk, Charles Messier spent his nights looking out for arguably the most important comet in history. In the back of his mind Charles was no doubt thinking about the spectacular sight of Comet de Chéseaux that he had witnessed in his youth.

Messier created charts (exquisitely decorated with the mythical figures of the constellations) using Delisle's calculations of the comet's path and scanned the area that Delisle had asked him to (and had marked on the map, which took into account the gravitational effects of large planets such as Saturn and Jupiter). Not being mathematically proficient, unlike Delisle and other colleagues, Charles likely didn't attempt such calculations himself. In return Delisle was probably glad to have his enthusiastic young depot clerk search in his stead.

Methodical, accurate, enthusiastic – Charles was all of these things, and the work must have been painstaking. Astronomy wasn't actually one of his employed duties, but he observed every clear night using a wide-field Newtonian reflector with a hefty 53-in. focal length. But one of Charles's favorite telescopes was a 6-in. aperture Gregorian reflector. Reflecting telescope mirrors of the time weren't made from precision, diffraction-limited glass, as they are today, but instead they were formed from the very reflective (but brittle) copper-tin alloy of speculum. Speculum was poor compared to modern aluminized glass mirrors and also had an annoying habit of tarnishing easily. But even considering this, the Cluny telescopes weren't of the greatest quality for their age. They must have given anyone using them (including Charles) quite a headache, not to mention eyestrain! Another curious fact about the Cluny instruments was that they had fixed magnifications. Thus they didn't have interchangeable eyepieces, and this necessitated a cumbersome array of instruments. But even with all of these telescopes at his disposal, try as he might, Charles just couldn't find the comet. Were the instruments just too plain awful to use…or was Halley's prediction wrong? (Fig. 1.5).

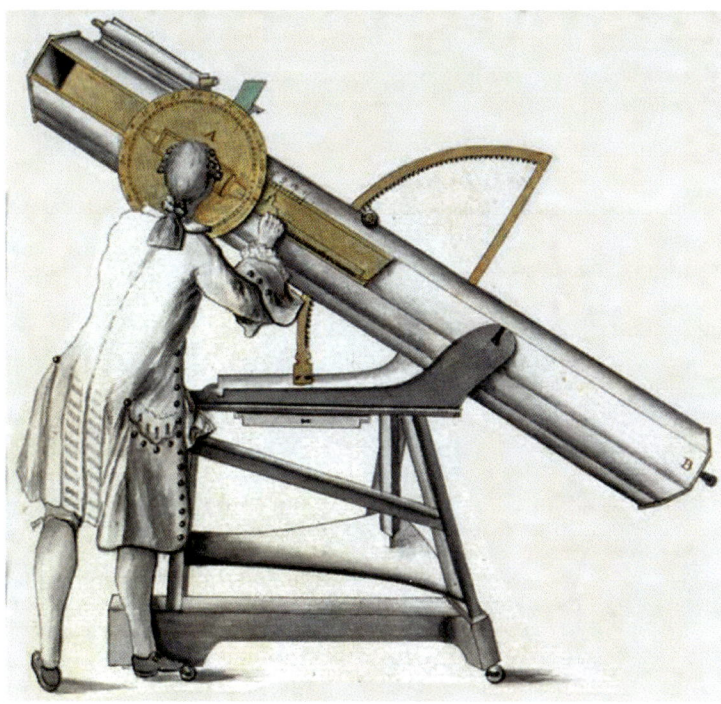

Fig. 1.5 Charles Messier using one of the Cluny telescopes, probably the 53-in. focal length reflector (Image courtesy of the Paris Observatory)

The answer to that question appeared to come just in time, for in 1757 Charles saw a fuzzy object in the sky (cometary halos often have a fuzzy appearance) in the constellation Andromeda. But as he looked more closely Charles could see that something was wrong. This object wasn't in the right place. Could it have been another comet? Charles observed for a few nights and found that it didn't move. Something that didn't move couldn't be a comet. It was in fact a tiresome impostor: a nebula. This object would eventually become M32 (or Messier 32) when Charles would later note it down in a list. Then on the 14th of August, Charles noticed what seemed to be a genuine comet in the sky. Had he finally found it? He checked the orbital path on his star chart with haste. His heart must have sunk a little when he quickly realized that the two didn't match. It was indeed a comet this time, but not the one predicted by Halley. And as a further blow, Charles couldn't even claim it as a brand new discovery.

The comet that Charles saw that evening was C/1758 K1 de la Nux, discovered on May 6 of that year, which, coincidentally, was the fifth anniversary of Charles's first observation at the observatory: that of the transit of Mercury across the Sun's face.

Despite the fact that it wasn't the comet he was after, Charles was nonetheless intrigued, and continued observing de la Nux for many nights. And then he spotted something else in the constellation Taurus on the 28th of August. Was it another comet? It would have to wait for the time being, as Charles continued observing de la Nux right up until 2nd of November, all the while keeping careful and accurate records in the way that Delisle and Libour had showed him. It was only after de la Nux disappeared from view that Charles returned to that curious fuzzy patch in Taurus. Had it moved? He saw that it hadn't, so it couldn't have been a comet. Charles noted the position of this nebula and labeled it Messier 1 (M1). It was this object that was to be the start of Charles's now-famous list.

Messier 1's position was at the exact same point in the sky that a bright new 'star' appeared seven centuries earlier. So bright was it at the time that it outshone Venus and was even visible in the daytime. What's more, Chinese astronomers of the time noted how you could read by the 'star's' light on moonless nights! But then the 'guest star' eventually faded over the weeks. Messier 1 is the remnant of an old, dead star that blew up in a supernova explosion, briefly outshining all the other stars in the galaxy put together. This is what gave it the appearance of a 'guest star' to the Chinese and others.

The fuzzy patch that Charles observed (first spotted by the English astronomer John Bevis in 1731) had now taken the place of the once-spectacular 'star.' In his time Charles wouldn't have known its true nature, but the third Earl of Rosse would name this patch the Crab Nebula in 1844 using his 36-in. reflector at Birr Castle in Ireland.

For Charles, however, the patch was just another frustrating 'non-comet,' much like the other fuzzy patch in the constellation Andromeda. Although he had noted it down using his own classification system he was really interested only in comets and saw these nebulae as a potential source of confusion and frustration. He would have to deal with these troublesome objects in a more systematic way at some point in the future. And in any case, the great comet hadn't yet returned (if it was going to at all), and that was the most pressing matter (Fig. 1.6).

Charles had now widened his search area beyond Delisle's remit and continued his systematic and relentless hunt. One can imagine that only daylight and cloud cover stopped him from watching the skies every waking moment. Then on the 21st of January 1759, two years after he started his search, Charles Messier saw something out of the corner of his eye with his wide-field Newtonian reflector. What was this new object? Could it be another dead end? After all, Charles had been searching for so long. Charles investigated the object further, comparing its movements closely against his charts. He checked and checked again, looking for some error that he may have made, or some divergence from the numbers. There was none. Charles saw excitedly that it matched Halley's criterion. And this object clearly wasn't a nebula, as it moved across the sky. It could only be a comet. It had to be *the* comet. Just 52 days before its perihelion (where it would have become entirely lost in the Sun's glare) the great comet had indeed appeared in the eyepiece of the wide-field Cluny reflector.

Finally Charles had found Edmund Halley's comet! But why did it take him so long? Charles quickly realized that Delisle's charts contained an error – an

The Race for Halley's Comet

Fig. 1.6 Halley's comet, the most famous comet in history, was the first to be shown to be periodic. Charles Messier, however, would have a difficult time with this comet (Image courtesy of NASA)

underestimation of the perturbing effect of Jupiter on the comet. This meant that Charles couldn't have hoped to be looking in the right place for the comet had he not used some initiative. But it had at last been found, right where Halley predicted. That surely was a cause for celebration regardless?

Actually, it seemed not. Delisle behaved strangely from that point on. He told Charles to continue observing, but would not accept that he had made a mistake. In addition Delisle also refused to make a public announcement about the return of the comet. Not wishing to upset his mentor and benefactor, Charles agreed to Delisle's somewhat odd request and continued as before, although it must have been an anguishing experience for him. The whole of the astronomical world had been waiting for the predicted return of the comet, and now after two years of sleepless nights and unrelenting persistence, Charles had been asked by Delisle (probably out of his own shame for the error) to keep it a secret.

In a time when discovering comets conferred the same celebrity status that reality television does today, this was indeed odd behavior on Delisle's part. It is all the more strange when you consider that back then, the exact masses for Jupiter and the other distant planets weren't known anyway, which is why the exact date of any comet's appearance couldn't have been predicted even to within months. Even skilled astronomical calculators like Lalande couldn't do that. In light of this, Delisle's apparent worry over his error seemed quite unnecessary (though the charts did indeed contain an error on Delisle's part, as well as this general uncertainty).

Then for Charles came the shocking hammer blow. It turned out that all his efforts were in vain anyway. Sometime in late March to April, news reached the Hôtel de Cluny that the comet had already been spotted by Johann Palitzch, an amateur astronomer in Saxony, Germany, on the night of December 25–26, 1758. The single-minded tenacity of a man who had spent Christmas night looking for the comet had paid off. Bitterly, Delisle's error had cost Charles the main prize. Palitzch had become an overnight sensation, and his achievement was a cause for celebration. It proved Halley's theory was true; comets were periodic after all. It was another resounding victory for science, and the comet of Christmas 1758 was officially named 'Halley's Comet' in the late Englishman's honor. Along with Palitzch, Halley's name was now renowned throughout Europe. Rather surprisingly, though, the comet's return was a somewhat tepid affair in Britain, Edmund Halley's own country.

Tepid must have also described the atmosphere in the Hôtel de Cluny. Delisle did eventually announce the return of Halley's Comet soon after the announcement from Saxony. But other astronomers in France were skeptical. Why, if the comet was discovered on the 21st of January, had the information been held for over three months? It was perhaps unfortunate that Delisle chose to make his announcement on the 1st of April.

If Charles was upset with Delisle over the whole affair there is no record that he ever said so at the time, though much later in his life, when writing his memoirs, he did express his regret and frustration that the great comet of 1759 had slipped away from him. But Charles, ever-tenacious, continued to work. In late January 1760 he discovered C/1760 A1, the great comet of that year (though it had also been spotted by many others) with a fantastic 5-degree long tail – 10 times the diameter of the full Moon in the sky. Delisle by this point had now become quite irrational and briefly also refused to publish this new discovery by Charles. In the end, though, the aging Delisle eventually changed his mind and allowed Charles to continue. The old man withdrew from astronomical work from then on, leaving Charles to dance to his own tune. One of his first observations after emancipation from Delisle's influence was his second Messier object (M2), a globular star cluster in Aquarius that the French-Italian astronomer Jean-Dominique Maraldi had discovered in 1731.

Clock Watching Beyond the Sea

Charles Messier had by now become France's human version of the Hubble Space Telescope, consistently finding new objects or observing others in detail. Mirroring his first-ever observation at the Hôtel de Cluny, Charles tracked a planetary transit across the Sun's disc. This time it was Venus, on the 6th of June, 1761. Saturn's rings were another observing target for Charles.

It wasn't long before he observed another comet: Klinkenberg (C/1762 K1 in 1762). Charles discovered Comet C/1763 S1 on the 28th of September, 1763, and C/1764 A1 on the 3rd of January, 1764. These objects didn't conform to any known cometary bodies, so they were totally new, and Charles Messier had been the first person on Earth to spot them. Probably buoyed by these, his first genuine discoveries (all the other objects having been discovered by others – not that it mattered as much in those times as it does today), Charles tried to become a fellow of the French Royal Academy of Sciences, the Académie des Sciences. The academy was founded by King Louis XIV in 1666 in order to procure the best intellectual talent from Renaissance Europe. One of the founder members was none other than the Dutch scientist Christiaan Huygens, known for his discoveries of the Martian polar ice caps and the rocky composition of the rings of Saturn, as well as the discovery of its largest moon, Titan. The probe that landed on Titan in January 2005, Huygens (part of the European Cassini-Huygens mission to Saturn), is named in his honor. Charles would have been proud to be counted among such hallowed figures, but he was to be disappointed when the academy rejected his application. Likely this was due to the fact that Charles wasn't an academic astronomer, not knowing astronomical theory, nor indeed being particularly skilled in mathematics. Charles himself knew that a lack of knowledge and experience in these fields held him back somewhat, but a rejection by the academy must have been a blow all the same. There must have been some part of him that felt the whole affair over Halley's Comet had its part to play as well.

Ever passionate though, Charles continued observing the skies, and curiously it was after his rejection by the academy that the number of discoveries he made really took off. By 1765 he had discovered a further 20 Messier objects. Many of these were star clusters; others were nebulae, galaxies, and stellar remnants. Though through a telescope some structure can be seen in a star cluster, there isn't much, visually speaking, to tell the other types of objects apart. The nature of galaxies as vast groupings of star systems was unknown until the early twentieth century. For this reason, all non-moving fuzzy celestial objects were called nebulae in Charles's time. The important thing about them as far as he was concerned was that they looked like, but were not, comets.

With the nebular objects Charles had decided (probably with his observation of M1) that he would finally catalogue as many of them as he could. He wasn't interested in them per se, but because their faint fuzzy appearance was so similar to that of comets he didn't want to confuse the two sets of objects any more – this source of irritation had to be dealt with once and for all, and the list of nebulae would become the world-famous Messier Catalogue.

Although at first it seems strange that a comet hunter would put so much effort into actively finding nebulae, you quickly realize that it is in fact a sensible thing to do. If Charles found and catalogued as many nebulae as possible, he would know exactly where they were in the sky because their positions would never change compared to the background stars. If Charles revisited a patch of sky and saw a new 'nebula,' then the chances were that it would be a comet (though watching the position of the comet changing from night to night would be the only way to be sure).

The cataloguing of nebulae then would be an invaluable method of avoiding false targets. There were other such catalogues available to Charles at the time, including those of Halley, Nicolas Lacaille, Giovanni Maraldi, and Johannes Hevelius. He incorporated the information from these into his own catalogue, but Charles wanted to expand on them to create a list as comprehensive as possible. He also wanted to check the previous observations for himself.

During this time Charles was also busy building up contacts with his contemporaries in Britain, Russia, and Germany. Clearly it seemed he felt recognition was deserved for his work, and joining an academy would be a great way of opening doors. Despite being shunned by his own academy in France, Charles was accepted as a member of the Academy of Sciences of the Netherlands (known as Harlem). On the 6th of December of that same year he was also accepted as a foreign member of Britain's prestigious Royal Society. The society was much more relaxed about admitting new members at the time than it eventually became.

A year later, in 1765, the 77-year-old Joseph Delisle, having taken a backseat for a few years, had now decided to retire from his post altogether. Though it took at least another six years for the depot clerk Charles to be appointed Astronomer of the Navy, he was now even freer to pursue his own observing program – and his unquenchable enthusiasm saw to it that he did exactly that. And 1765 was significant also because Charles discovered his 41st Messier object, an open star cluster in Canis Major. Like most of Charles's other discoveries this one had been discovered before. In this case the Sicilian Giovanni Hodierna saw it at least 112 years previously, and the ancient Greeks probably knew about it too. Nevertheless, being the first to the post wasn't really the point of Charles's catalogue of nebulae; he simply wanted to avoid confusing these objects with comets.

It was on the 8th March in 1766 That Charles Messier, who was becoming renowned in Europe as a prodigious comet hunter, discovered another one of these much sought-after celestial visitors (C/1766 E1). This time he spotted it with the naked eye, which is how he also discovered the great comet of 1760. This was probably just as well, considering the quality of the instruments he had to work with. In April he independently co-discovered Comet D/1766 G1 along with Johann Helfenzreider.

It was in 1767 that Charles, who after Delisle's retirement was at this stage still only a naval clerk, undertook his first (and only) ever-recorded naval voyage. Along with a colleague Charles was to test marine chronometers made by Julian and Pierre Le Roy, who were the French equivalents of John and William Harrison. John Harrison was the Yorkshire-born carpenter's son famed for creating a series of increasingly accurate marine chronometers to aid naval vessels in their calculation of longitude. Despite Nevil Maskelyne, Sir Isaac Newton, and other scientific heavyweights championing the lunar table method, longitude calculations at sea could only really be done in any practical way with the aid of an accurate timepiece. Though pendulum clocks existed, rough sea voyages jostled ships up, down, left, and right by just the right frequency and amplitude to play havoc with the mechanisms. Three years before Charles's own voyage, Harrison's H4 model was successfully tested on a return voyage to Barbados. Harrison's tale is recounted in Dava Sobel's *Longitude*.

Le Roy, like Harrison, was also working on the problem of longitude and had invented a bimetallic strip to compensate for the changes in temperature during a voyage (particularly when traveling from cold, northern European waters to the tropics, and vice versa). Normally, differences in temperature would make the metal in any chronometer expand and contract enough to alter its timekeeping abilities, but the bimetallic strip would overcome this. Harrison had also arrived at the same solution independently. The question of accurate timekeeping was arguably the most pressing technological problem in the age of the European trading empires – the 'golden age' of maritime voyages. Whoever accurately knew the time would effectively have mastery of the sea, and hence the world's naval trading routes. Prophetically, the British had gotten there first.

But France wasn't going to give up so easily. Le Roy's latest chronometer was considered to be equivalent in prowess to the H4. A ship known as L'Aurore (the Dawn) was built in La Havre docks on the Normandy coast specifically for the purpose. Both the ship and the voyage were funded by the Marquis de Courtanvaux. L'Aurore is not to be confused with another vessel of the same name that served as a slave ship, transporting people from Angola to the West Indies between 1784 – when that ship was built – and 1789 when the slave trade was abolished in France (albeit briefly).

Charles's colleague on his 3–4 month voyage was Alexandre-Guy Pingré. Unlike the H4 on its voyage, the Le Roy chronometers and the two accompanying astronomers would not be traveling to the warm paradise islands of the tropics. Charles and Pingré's journey would instead take them to the colder waters of the Baltic Sea in Scandinavia. In order to show that the marine watches were accurate it was necessary to test them against the tried and trusted sextant (used since 1731), despite the formidable difficulties of using one at sea. But if anyone could do it, the exacting Charles Messier could. Charles and Pingré formed a two-man team, with Charles doing the astronomical observational work and Pingré doing the calculations. During Charles's absence from the Cluny observatory, Lalande continued with the observation program.

Charles it seems didn't really enjoy his excursion (important as it was), and was probably glad of returning to France four months later.

Recognition, Tragedy and a Big Egg

In April 1769, Charles was elected member of the Stockholm Royal Academy of Sciences. Then on the 8th of August, Charles discovered the great comet of that year, which, like his previous discoveries was named after him. This latest one was called 1769 Messier. The modern-day designation is C/1769 P1. Being shrewd, and no doubt remembering the debacle of Halley's Comet (as well as the snub by his own academy), Charles sent a letter and a diagram of the comet's position to Frederick II, King of Prussia (a now defunct kingdom of the German Empire). Reputedly it is said that Frederick was so impressed with Charles that he pulled a few strings and got him accepted as a member of the Berlin Academy of Sciences on the 14th of September.

And the discoveries just kept coming, with yet another comet for the ambitious and determined Charles, spotting it as he did on the 14th of June 1770. 'Messier's Comet' as it was called (in the same vein as Halley's Comet) was not to be named that for very long. That honor would go to Anders Lexell, a Finnish mathematical astronomer on placement at the Observatory of St. Petersburg. Lexell calculated the path and shape of this comet with a period of 5.6 years. This was an astonishingly short period and a virtually unheard of thing at that time. The comet that Charles discovered was then officially named Comet Lexell, though to this day it has never been seen again. Lexell's Comet was probably fatally captured by Jupiter, or, the giant planet flung it out of the Solar System altogether. To this day, no one knows the comet's ultimate fate.

Then in that same year Charles's position was finally vindicated. He was accepted into the Académie des Sciences on the 30th of June as an 'academian,' much to his delight. Finally it seemed his hard work and tenacity had earned him the respect of his fellow scientists and countrymen. And more importantly, his cometary work was now widely recognized. The honorable academy members could resist the pressure no more to duly reward this man for his efforts. The six-tailed comet of de Chéseaux that he had once gawped at with amazement in his youth now must have taken on a great significance in Charles's mind, like a vivid, congratulatory memory. Surely also the bitter memories of Halley's Comet must have diminished greatly. The news of Charles's membership even reached King Louis XV of France, who in a well-natured manner joked that Charles was, "the 'bird-nester' of comets!" Charles Messier could have wished for no better royal endorsement.

And he was to have an even happier experience later that year. Charles had been acquainted with Mademoiselle Marie-Françoise de Vermauchampt for over 15 years. Like himself, she was from Lorraine. Europe's greatest comet-hunter summoned the courage to ask Mademoiselle Vermauchampt for her hand in marriage, and so on the 26th of November 1770, they were wed.

By 1771 Charles Messier had listed 45 nebulae and so now his first catalogue was published in the Academy's *Memoirs*, with Lalande kindly writing an introduction for him. The naval clerk from Lorraine who had started his career copying out maps by hand was now promoted to the position of France's Astronomer of the Navy and his salary was increased, first to 1,700 francs, and then to 2,000 francs several years later.

Everything was looking up for Charles. Marie-Françoise and he appeared to be a devoted couple, but the marriage was to be tragically short. The following year on the 15th of March, Marie-Françoise gave birth to a son, Antoine-Charles. But Marie-Françoise developed complications during the birth that she wasn't able to recover from. She died on the 23rd of March, and as if by some horrible design, little Antoine-Charles followed too, three days later. So just like Charles's siblings and father, his wife and son had been taken away from him. Charles, though having a flourishing tree of success had to, it seemed, also to accept its bitter fruits. The exact cause of Marie-Françoise and Antoine-Charles deaths aren't known, but these events are recounted in more detail in *Le Furet des Comètes* by Jean-Paul Philbert (French only).

It was remarked by some at the time that Charles grieved more for the loss of another potential comet discovery by having to be by his wife's side, than for her passing. There is no evidence to support this and it's likely to be a cruel apocrypha born of a dark joke.

Life had to go on for Charles and his discoveries continued too, with more comets in 1772 and 1774. It was in this year that a new face appeared at Cluny. A student of Lalande's, Pierre Méchain became 'calculator' at the Depot of the Navy. Like Charles, Méchain was a keen comet hunter and spent his spare time at Cluny hunting for comets. What's more, unlike Charles, he was a good mathematician. This meant that Méchain could calculate the predicted orbits of comets himself without having to rely on the skill of others as Charles had to. Did Charles Messier now have a rival in his midst? In a way, he did. But no documented evidence suggests that Charles Messier and Pierre Méchain were anything other than firm friends, and the competition between them good-natured. And in any case, Méchain only discovered his first two comets two years later.

Despite that, Méchain seems to have been blessed as an observer, for one night at the observatory he hit the 'mother lode'. Under Charles's guidance Pierre Méchain discovered a whole raft of nebulae in the Coma Berenices/Virgo constellation region. He had in fact found what we know today as the Virgo Supercluster – the most galaxy-rich region in the night sky and the largest grouping of galaxies in the Universe to influence the Local Group, to which our Milky Way Galaxy belongs. The Virgo Supercluster would be the eventual location of the 100th Messier object. Though friendly rivals when it came to comet hunting (even considering that two determined men as these must have got quite competitive) Charles and Pierre co-operated when it came to the nebular objects. Méchain would report his observations to Charles, who would then make a record of them for a future, additional catalogue. With the addition of Méchain's contributions Charles would publish his enlarged catalogue in 1781 in the French ephemerides *La Connoissance des Temps* for 1784. When it was published there were now 103 fuzzy nebulae in the Messier Catalogue (Fig. 1.7).

The last three objects (M101, M102 and M103), which were observed by Méchain were a last-minute rush-job, stuffed into the catalogue just before publication – Charles had no time to double-check them. In the meantime, with Méchain's help, Charles's own list of objects continued to grow ever more.

That same year a German-born British musician, Bath Orchestra Director and stargazer called William Herschel was observing the skies with a homemade telescope of his. Along with his sister Caroline the pair were known for making the finest-quality speculum mirrors around, even though music and not astronomy was their actual profession. And yet the Herschels' skill in telescope making was to prove advantageous on the 13th of March when William Herschel spotted what he initially believed to be a comet. On hearing about this new discovery on the 4th of April through the British Astronomer Royal Nevil Maskelyne, Charles stopped his usual searches in order to observe it. He passed on his observations of 'Herschel's

Fig. 1.7 An elderly Pierre Méchain, shown wearing the cross of the Legion of Honor (Image courtesy of the Paris Observatory)

comet' to his friend: the astronomer, mathematician and Judicial President of the French Parliament, Jean-Baptiste-Gaspard (or 'Bouchart') de Saron. Charles wrote a glowing letter to Herschel, saying at one point, "It does you the more honor as nothing could be more difficult than to recognize it and I cannot conceive how you were able to observe it several days in succession to perceive that it had motion, since it had none of the usual characteristics of a comet." Herschel quite likely could see the object much better with instruments that were far superior to those at Charles's disposal. But this last point made by Charles was to prove telling. As he had detected a small degree of movement in this object, it certainly wasn't a nebula, nor even for that matter, a dim star. So was it a comet after all then? Lalande and Méchain had taken a look at Charles's observational data for the object. Using mathematical calculations, both they and de Saron had all independently come to the same conclusion: this was not a comet at all, but something much more astonishing. Their figures showed that Herschel's object was much too far from the Sun to be a comet, beyond even the orbit of Saturn. Only one conclusion could be drawn for something so far away that was still visible: this had to be a new planet! Surprisingly perhaps, it had been discovered before on numerous occasions, including by none other than John Flamsteed in 1690. However, it had been dismissed by him, as well as other astronomers, as simply a star. Not so by the methodical Charles. With the help of his mathematically proficient colleagues he was able to show with his careful observations that Herschel had discovered a new world of the Solar System. Even though it was on someone else's behalf, the 'bird-nester' had helped bag the biggest egg of them all. Other astronomers in Europe soon confirmed what they too initially suspected (but never followed up) and William Herschel became a celebrity. He had wanted to name the new world 'Georgium Sidus' after King George III gave him a £200 annual pension and became his patron, though it would eventually be named Uranus, in keeping with the classical tradition of naming planets. Herschel though wasn't to forget Charles's contribution.

Meanwhile, the coming November heralded the second time in Charles's life that he would befall a nasty accident. This time it was to be the misfortune of a grown man rather than the boisterous activities of a child. Whilst out walking in the grounds of the Duke of Orleans with his friend de Saron, Charles went in through an unlit doorway, not realizing that it was an ice cellar. The unfortunate Charles fell 25 feet into the cellar, breaking his leg, wrist and several ribs. Charles was rescued from his awful fall but he was now virtually bedridden. In that time he must have thought constantly about his nebulae and all the potential comet discoveries he must be missing. But he could do little about it, except heal as quickly as possible. In the end it took a full year before Charles was up and walking again. Though he was up and about by the 9th of November 1782 Charles would walk with a limp for the rest of his days.

It was during 1782 that Pierre Méchain managed to be accepted by the Académie des Sciences, seemingly with much more ease than Charles ever did. There was a good reason for this. Méchain had discovered two new comets and calculated their paths. He had also used mathematics to destroy the notion that the appearance of a comet in 1532 and in 1661 related to the same object. Though flushed with success Méchain must have been pleased by the eventual recovery of his friend Charles, and no doubt relished the continuation of the friendly competition between the two.

Revolution and Napoleon

In the period after his fall, the first thing that Charles observed was another transit of the planet Mercury, on the 12th of November. At one time some years before, he had entertained the idea that there was an unseen planet inside Mercury's orbit, though he never found any real evidence for it, and it appears that he quickly dropped this line of investigation. From 1782 onwards Charles didn't work on his catalogue much more (apart from some tidying up), as William Herschel had started a corrected version at some point in the future. In the meantime instead of concentrating on his catalogue he used this time for his first love: discoveries. Méchain though continued with the Messier catalogue, discovering objects M105, M106 and M107 in 1783. He had also discounted M102 (one of the three objects Charles didn't have time to check), as he believed it to be a duplicate observation of M101.

In 1784 Charles observed the great comet of that year, discovered by Cassini on the 24th of January. Charles was informed of this the day after and was encouraged to follow it, but cloud cover meant he couldn't do so until the 3rd of February. Charles followed it with the 12-in. focal length refractor at Cluny until the 25th of May, where it disappeared from his view past Andromeda. The next year Méchain discovered two comets (one of those jointly with Charles) and Charles also discovered one. Méchain would discover one more in 1787, and Charles another one the next year on the 25th of November 1788.

By now, Louis XVI had been on the throne of France for 14 years. His late grandfather Louis XV, the one who had dubbed Charles the 'bird-nester of comets', and who was an intelligent and private king, nevertheless had spent vast sums of money on wars with France's neighbors (though many of them were victorious). One of these was the Austrian War of Succession, which had touched Charles's own home region of Lorraine. Though that war too was eventually won, Louis XVI had now inherited a country with a huge national debt and a tax system that heavily favored the wealthy and punished the poor. Starvation was spreading through parts of France due to the high price of wheat, which was not helped by the imposition of a further tax on the crop (called the 'dime') by the Church. Grievances against the 'Ancien Régime' (Ancient Regime – the system of feudal monarchy) were growing. Louis XVI – intelligent though he was – was an insipid king; unable to be decisive about anything (though he did manage to send naval fleets to the American colonies to support them in a war being waged there with Britain for independence). On top of his ineptitude and his country's disastrous economy, Louis even fired his finance minister for daring to suggest that he cut back on his opulent lifestyle. For the people of Paris, this was the 'straw that broke the camel's back'.

Just over a kilometer away from the Hôtel de Cluny across the River Seine, the Bastille Saint-Antoine (a small prison) was stormed and broken into on the 14th of July 1789, in order that arms and ammunition stored there be accessed. This is probably the most famous event that marked the beginning of the French Revolution. Charles Messier, ever the tenacious and single-minded astronomer, gives the general impression that the revolutionary turmoil of the next decade was simply an irritating hindrance to his work, though in reality he was to suffer personally. Though his own position as Astronomer of the Navy ceased to exist, along with his salary, Charles's simple complaint to the navy was that he did not have enough provisions, oil or candles to light and heat the observatory so he could continue working. He found instead that he had to share with Lalande.

It was on the 27th of September 1793 that Charles observed his 36th comet, and this one was another original discovery. Even after the Storming of the Bastille, up to this point in time Charles continued making observations of comets, either discovered by himself or by others. Charles would announce this latest discovery of his to Lalande, and he says at the time how the discovery was announced to newspapers also.

But if Charles was to open any Parisian newspaper at that time, he would have seen that the revolution in France wasn't abating, and had in fact got much worse. The Reign of Terror, or 'la Terreur' had started with rival factions of the Revolution vying for control. Louis XVI, grandson of the man who had honored Charles had already been executed by guillotine in January 1793, and in October the Queen, Marie Antoinette, would suffer the same fate.

Charles would continue observing his September comet up until the 11th of October, all the while determining its position. He sent his observations to de Saron so that he could calculate its orbital parameters. Bouchart de Saron was grateful to Charles for his request, for as a former president of the now-dissolved parliament he had been imprisoned as an enemy of the Revolution. De Saron was was glad to have anything to focus his mind upon, seeing as how he was awaiting execution. Charles used his friend's calculations to successfully track down the comet, and was able to let de Saron know this just before he was led to the guillotine on the 20th of April 1794.

Pierre Méchain could not escape the effects of the Revolution either, even though he was far away from Paris at the time. Both he and Joseph Delambre were appointed to map the line that would form the Paris Meridian for the purposes of longitude (the Greenwich Meridian was chosen in the end as a matter of history) using techniques similar to that of the Romans when they built their straight roads. The chemist Antoine Lavoisier, who put together the funding for the project, also thought out the actual measuring technique. Delambre would start in Dunkirk, Méchain in Barcelona, and the two would meet halfway. Delambre had encountered significant problems doing the work due to the political turmoil in France, and Méchain was even arrested as an 'anti-revolutionary spy'. His equipment at first was not understood and was viewed with suspicion. However, he was eventually allowed to continue, but then he had an accident in Spain and was forced to remain there. When he had recovered from his injury Méchain was held back again. A war had erupted between Spain and France, no less, and Méchain, in a tragic-comic fashion, was imprisoned for a second time. The survey was finally completed in 1798, six years after it had started. Méchain's family, however, had lost all of their land and possessions in the Revolution. And Lavoisier as a taxman, was an enemy of the Revolution. He had met his end at the guillotine some four years before (Fig. 1.8).

Rather fittingly, the architect of la Terreur, Robespierre, was himself guillotined on the 27th of July 1794. France breathed a sigh of relief and could now return to some semblance of normality. And just as the new France now had to adapt to being without a monarchy, Charles Messier had to adapt to life without a salary, a position, a pension – or indeed means of any kind. But he was lucky in his friends, who were only too happy to help him out, especially the generous Lalande.

Despite all their setbacks and misfortunes, Charles and Méchain still continued to find comets in the most turbulent of times. Truly, their observing skills were formidable. After the Revolution they were both awarded membership of the newly formed Academy of Sciences (now with the omission of the word 'Royal') and also, membership of the French Board of Longitudes. And for Charles himself his irrepressible efforts were to be further recognized by no one less than the most famous (or infamous) general of the Revolution himself – and one the most successful military leaders in history: Napoleon Bonaparte. Napoleon awarded Charles the Cross of

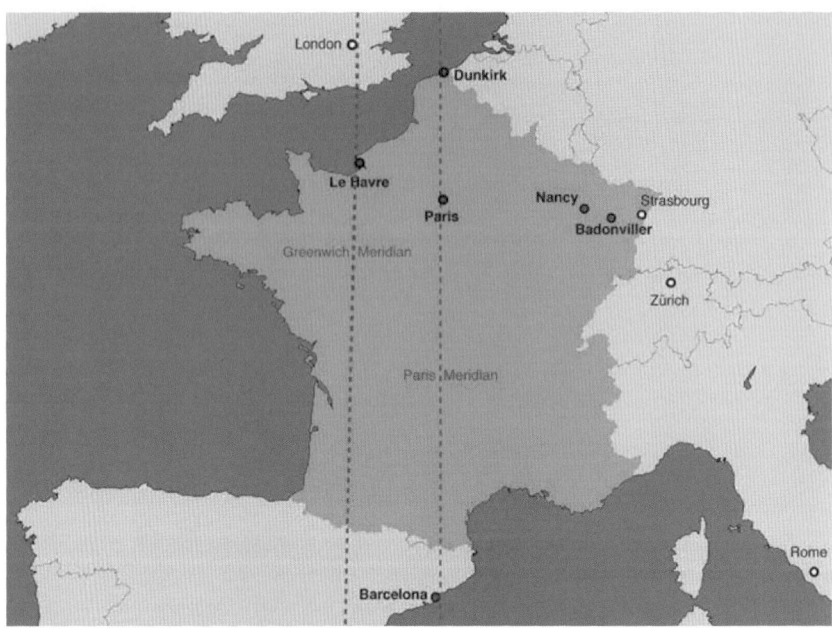

Fig. 1.8 Map showing the Greenwich and Paris meridian lines. The Paris line is effectively the route taken by Delambre and Méchain starting at Dunkirk and Barcelona, respectively (Image by Kulvinder Singh Chadha)

the Legion of Honor with his own hand. So moved was Charles by this gesture that he later wrote a gushing memoir, describing Napoleon as 'grand' and honoring him with the comet of 1769 (the one that Charles had written to Frederick II about), the year of Napoleon's birth. Charles's oddly astrological sycophancy was probably forgivable in the eyes of his friends and colleagues given the fact that the hardworking comet hunter was now 78 years of age. Probably Charles felt that he had finally got the recognition that he richly deserved for his work (something that Herschel and the mathematician Jean le Rond d'Alembert stated as well). But Napoleon was a famously impenetrable and complex figure and there's no evidence that he ever responded to Charles's gesture anyway.

Although Charles continued to observe as much as he could, he was nearing retirement. Crucially he never got round to republishing his catalogue in order of right ascension, nor did he ever publish an additional catalogue with the extra observations, as he intended to do at some stage. Though the Revolution was over, France was still at war (particularly with Britain) and there was no money now to repair the increasingly dilapidated observatory of the Hôtel de Cluny. Charles Messier would never have it refurbished. He suffered a stroke in 1815 and died on the 11th of April two years later. The 'bird-nester' would bag no more comets (Fig. 1.9).

Fig. 1.9 An unusually young looking Charles Messier sporting the cross of the Legion of Honor, presented to him by Napoleon (Image courtesy of the Paris Observatory)

The Legacy of Charles Messier

In 1775 during Charles Messier's lifetime, Lalande proposed that a new constellation be named in the man's honor: Custos Messium, 'the Harvest Keeper'. Charles was to be immortalized in the group of stars that lay between Cepheus, Camelopardalis and Cassiopeia. Though included in some star atlases of the time including Bode's *Uranographia*, and John Flamsteed's French version of his star atlases, Custos Messium wasn't to last. Though he was being sincere about his friend, Lalande quite often proposed new constellations (often when inebriated), usually in the shape of cats and balloons. A more lasting tribute to the man has however survived in the form of the Moon craters Messier A and Messier B, situated in Mare Fecunditatis. But perhaps Charles Messier is best remembered for the catalogue that bears his name even though in his lifetime he observed and tracked 44 comets. Thirteen of these were his own original discoveries, and a further seven were independent co-discoveries. That's almost half of all comets discovered in Europe at the time connected in some way to a single individual.

With help from Pierre Méchain, Charles catalogued a total of 103 nebulae between 1757 and 1783. A further seven were rediscovered in the 20th century by Dr Helen Hogg of Dunlap Observatory (University of Toronto), and Dr Owen Gingerich of Harvard University. Pierre Méchain's own discounted observation of M102 was also re-included. Instead of a duplication of M101, astronomers today believe that M102 is in fact the Spindle Galaxy in Draco. So in total, Messier's catalogue today contains 110 objects, from galaxies, nebulae and star clusters to supernova remnants. A nice, even number of objects to depict in a pictorial grid on a poster.

Charles Messier's methodical approach to his work paid off time and time again. William Herschel was inspired by Charles's catalogue, and was no doubt grateful for his critical contribution in the discovery of Uranus. Herschel began his own catalogue of nebulae, and with his superior instruments found over 1,000 fixed fuzzy objects. The Herschel catalogue quickly expanded to over 2,500, though not all of the objects were real nebulae and the total number was reduced somewhat. The reason that Charles had stopped searching for nebulae after returning from his accident to Cluny was because he knew that the Herschel Catalogue would supersede his own. And just as Herschel was inspired by the Messier Catalogue, the 20th century Dutch-Irish astronomer Johann (later John) L. Dreyer would be inspired by the Herschel Catalogue. Dreyer formed the most comprehensive deep sky list ever: the New General Catalogue (NGC), used to this day by observational astronomers, academics and space agencies such as NASA and ESA. Although the New General Catalogue is the definitive list of deep sky objects, presently containing nearly 8,000 entries, there is a strong tendency amongst amateur astronomers to refer to Charles's objects by their Messier numbers – such is the fondness for the Messier Catalogue. If Charles were alive today he'd surely have been amazed to see modern amateur astronomers follow in his footsteps, in an accelerated mode, by challenging themselves to finding all 110 objects in one night. This feat Charles himself no doubt would have loved to accomplish, as it would have meant more time for comet hunting. In that regard, would Charles have delighted in the modern 'Messier marathon' (as the observing challenge is called) or would he have been dismayed at being remembered more for his catalogue instead of for his comets? Charles Messier had a need of recognition, illustrated by his letter to Frederick II, his gushing note to Napoleon, and the fact that he had joined nearly a dozen European Academies in his lifetime. Taking those facts into consideration one would hope that this quiet man would be immensely proud to still be inspiring astronomers two centuries after his death.

Chapter 2

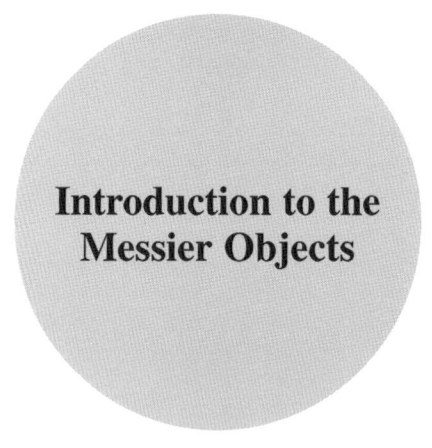

Introduction to the Messier Objects

From an amateur perspective, the Messier Catalogue contains many objects. Not only are they of different types, encompassing star clusters, galaxies and nebulae but also they are of different degrees of difficulty from those readily visible to the unaided eye to those that are tough with medium-sized amateur telescopes. Often, when asked the question "have you seen M81?", it is more accurate to reply that you've seen a fuzzy patch where M81 is supposed to be. As Charles Messier predates Edwin Hubble, he was unaware of the nature of many of the objects in his catalogue. That doesn't diminish his achievements in any way, as it is a good general list for deep sky viewing but it also succeeds in its original objective and many people (including me), resisted the temptation to go blabbing about a "new comet" after checking catalogue objects first.

Whilst most of the objects are not easily visible to small binoculars, often found in many households, there are certainly enough to be worth searching for. Some, like the Pleiades (M45) and Beehive (M44) are a splendid sight, in very modest binoculars, whilst others are best seen as objects to tick off the list of things to see before you die, rather like Uranus, Neptune and Pluto.

It would be a sweeping generalization to say that deep sky viewing is more difficult than solar system viewing, although they have different challenges and it's a matter of personal opinion as to what you find more difficult.

It would be true to say that most Messier objects are best enjoyed with an 8" or larger reflector with a short focal length and some sort of "wizardry" to find the objects automatically. However, such equipment was not available to Charles Messier and is certainly not available to me on a daily basis. Whilst he had a 6" Newtonian reflector, I have a Skywatcher 5" Maksutov-Cassegrain, which has slightly less light gathering power but sharper views. However, apart from the modern light-polluted skies of southern England, I also had the disadvantage of having an instrument with a long focal length. The work-round was to use a Skywatcher 32 mm focal length Plössl eyepiece and an Antares screw-in focal reducer, which reduces the magnification by 2× and increases the field of view by about 1.7×. The combination of both gives a magnification of 24× and a field of view of about 1.8°. A Skywatcher 9×50 finderscope makes finding objects easier, too. Using a light pollution reduction (LPR) filter may not sound quite in the spirit of the book but many of the Messier objects are barely visible without one, whilst for many it enhances the view quite considerably.

The other instruments used for observing the Messier Catalog are:

- Helios Stellar 15×70 binoculars
- Skywatcher StarTravel 80 short tube refractor

These give wider fields of view than the Maksutov-Cassegrain and using the 32 mm eyepiece and focal reducer combination achieve a whopping great field of view of 7° but only 6× magnification.

In general, the easiest objects to see are open star clusters, as they are less affected by moonlight and light pollution, the hardest are usually galaxies and globular star clusters, especially those whose light is spread out over a large area. Nebulae are almost as difficult but usually there is some sort of shape or internal contrast that makes them stand out from the background. Through modest equipment from difficult locations, it is actually impossible to tell a globular star cluster from a galaxy, as they can both appear as round fuzzy patches, with some central condensation of brightness. Indeed, in Messier's day, it is likely that few astronomers knew that they were different objects, with globular star clusters in and around our own Milky Way Galaxy and the others galaxies or "island universes" in their own right, many much larger and more massive than our own.

Before embarking on this project, most of my deep sky observing was limited to short sessions with binoculars, often just before bedtime and photographing double stars. This is partly due to a busy life with work and family but also because I live in medium light polluted skies, where only the brighter objects are visible and solar/lunar viewing is a lot easier. I had, however, explored the Ring Nebula (M57), Dumbbell (M27) and Andromeda Galaxy (M31) quite extensively with the Maksutov-Cassegrain.

The Messier Objects

This section contains descriptions and photographs of the Messier objects in catalogue order. The actual appearance of each object will vary according to:

- The clarity of the sky, affected by cloud and/or light pollution
- Eyesight quality of the observer
- Experience of the observer
- Moonlight

- Quality of the individual instrument and accessories used
- Extinction (dimming of objects near the horizon)

Extinction is a particular problem for many members of the Messier Catalogue, as they have a very southern declination. Indeed M7 is never more than 6° above the horizon anywhere in the UK and for the northern parts, never rises at all. As extinction becomes significant for objects within 15° of the horizon, many of the objects are never seen in their full glory from the UK (or even France for that matter). However, if you make leisure or pleasure visits to the southern hemisphere, many of them can be seen at their best. However, you may need to sacrifice aperture for portability.

The easiest objects in the Messier Catalogue are usually the open star clusters. Under poor viewing conditions, their fainter members are lost but, especially for the brighter clusters, their main asterisms are visible under conditions that you would not expect. For a minority of them, however, clear nights make them harder. Some of them are in the plane of the Milky Way, so they can be hard to pick out against the stellar background. Lose a bit of visibility and, hey presto, you can see them. For globular star clusters, galaxies and nebulae, their brightness is more evenly spread out, often over a large area. For example, the Andromeda Galaxy (M31) is the largest natural object that can be seen in the night sky from the northern hemisphere, having about six times the area of the Sun or Moon. This makes the published magnitudes a bit misleading when it comes to visibility. It is based on the total brightness of the whole object. As a general rule, if you can see say a 9th magnitude star through binoculars on a given night, you should be able to see a 7th magnitude galaxy at similar elevation from the horizon. So drop a couple of magnitudes off of star visibility and you're thereabouts. Indeed, for many objects, the published magnitude varies amongst the various commonly used sources and for the larger objects doesn't give a realistic assessment of how easily visible they are. Where there's wide discrepancies amongst various sources as to the magnitude of an object, I have taken an approximate average.

The faintest members of the Messier Catalogue are about magnitude 10.0 and large in size, so are very close to the limit of my equipment under very good conditions. Amazingly enough, there are some brighter objects of reasonable declination (as seen from the northern hemisphere) that he missed.

Quite often, the more difficult objects lack the "wow" factor that you get with some of the easier objects. Figure 2.1 shows the Andromeda Galaxy (M31) photographed through a small telescope and compact digital camera. The smudge in the middle looks nothing like it does through better telescopes and/or photographic equipment. Yet, this is how the more difficult ones look and the challenge is to spot them at all, although others can be a nice surprise. For example, M35, the bright open star cluster in Gemini proved to be a very pleasant surprise when viewed though my set-up, even though I'd seen it in binoculars many times before.

There's a section for each object, which gives:

- An overall description
- How to find it, including a diagram or reference to another diagram

Fig. 2.1 M31 by Anthony Glover

- A description of how it looks, mostly through my binoculars and Maksutov-Cassegrain
- Charles Messier's original observing notes
- Photographic details, showing how the object looks at its best and a representation of how it looks through my Maksutov-Cassegrain. Where applicable, I have included some simple photographs taken through my modest equipment, just to show it can be done.

Below is the introduction to Messier's notes, as researched by Kulvinder Singh Chadha:

Catalogue of Nebulae and Star Clusters

Observed in Paris, by M. Messier, the Naval Observatory, Hôtel de Cluny, Rue des Mathurin.

[This is the *Catalogue des Nébuleuses et des Amas d'Étoiles,* Messier's complete catalogue of his nebulae, which was included in the *Connaissance de Temps* for 1784 and published in 1781].

[As this is a translation, the grammar and sentence structure (e.g. that of many clauses) reflects the style of the time].

[The given size of a telescope refers to the instrument's focal length].

[Messier gives his Right Ascension values in degrees, minutes and seconds of arc, as well as hours, minutes and seconds. The diameter of each object (where applicable) is also given in degrees and minutes of arc, mirroring the columnated table in the published catalogue. Both his and Méchain's RA and Dec calculations for each object will inevitably differ from modern-day values. When looking for the Messier objects it is those modern coordinates that should be referred to].

M. Messier has observed with great care the nebulae & star cluster discoveries that can be seen over the skyline of Paris, he has determined their right ascension and declination, and gives their diameters, with circumstantial details of each: a work that has been wanting in astronomy.

It also enters into some of the details on the research he has made of nebulae that were discoveries by several different astronomers, but he has sought them in vain.

The C*atalogue of Nebulae and Star Clusters,* by M. Messier, is in the volume of the *Academy of Sciences, year 1771, page 435.* He reported at the end of his memoir a drawing traced with the greatest of care of the beautiful nebula in Orion's sword, with the stars it contains. His design might be used to acknowledge it again, if in the course of time it is not subject to any change. A comparison of this present design with those of Messrs. Huygens, Picard, de Mairan and le Gentil, it is surprising to find such a change that one would scarcely imagine that it was the same nebula, if one regards its form. You can see these drawings, given by M. le Gentil in the volume of the *Academy [of sciences], 1759, page 470, plate XXI.*

The catalogue M. Messier prints, which we give here, we report many nebulae and star clusters that have been discovered since the printing of his memoir, which we have communicated. (NOTE: Objects up to and including M45 are from his printed catalogue and objects from M46 to M103 are taken from his memoirs).

For the positions of the nebula, Messier refers to the numbers that are the same on the next page, and which give the details of each of the observed nebulae.

Chapter 3

M1–M22

M1

This is also known as the Crab Nebula. It was formed when a massive star exploded as a supernova. To modest instruments it doesn't really look any different from any other type of nebula but research pictures show an intriguing "butterfly" shape. It is actually six light years across and all of 6,500 light years away, giving an apparent size of 6×4 arcminutes, so is a comparatively large object, much larger in apparent size than any planet. It was discovered in 1731 by John Bevis.

The supernova apparently happened in 1054 AD, as seen from Earth as a bright "new star" but, due to the time light takes to reach us, it actually occurred over 7,000 years ago. It is the closest and easiest object of its kind and is still expanding.

As it is near the ecliptic, it is possible for solar system objects to pass close to it and even cross it but will not be visible if a bright object such as the Moon or Venus is nearby. Mars passed close by in September 2007.

Where to Find

M1 is quite easy to find, being about a degree north of the most southerly of Taurus's horns. Failure to see it is usually because of thin cloud, extinction or moonlight, rather than poor optics or looking in the wrong place (Fig. 3.2).

Fig. 3.1 M1 by Anthony Glover

Object Type	Nebula: Supernova Remnant
Declination	22° 01 min
Right Ascension	5 h 34.5 min
Magnitude	8.4

What It Looks like

Although it is the first object in the Messier Catalogue, it is actually one of the hardest, with a visual magnitude of 8.4, spread out over 24 square arcminutes. To modest instruments, it looks like a misty patch, with no sort of structure. Although it can be found in large binoculars on a good night, a 127 mm Maksutov-Cassegrain does little to improve the view.

Charles Messier's Original Notes

12 September 1758
(RA: 05 h 20 m 02 s, 80d 00′ 33″; Dec: +21d 45′ 27″; Diameter: –)

Nebula over the southern horn of Taurus, contains no star, but the light is whitish and tapered, like a candle flame, discovered whilst observing the comet of 1758. See the chart of this comet in *Memoirs of the Academy for the year 1759, page 188*. Observed by Dr Bevis in 1731, it is reported in the *English Celestial Atlas*.

How to View

Although large binoculars will reveal its existence from suburban locations, it will not show well and, simply, the larger aperture and clearer sky you can get, the better. Through telescopes, mid-range magnification of say 50–100x, gives best results and will not over-fill the field of view.

Photographic Details

M1 was taken using 6×15 min exposures with an 110 mm/F7 APO refractor, SXV H9, dark subtracted, stacked and processed in *Nebulosity*. A 12 nm hydrogen alpha filter was used, which is different to the narrowband ones used for solar viewing (Figs. 3.1 and 3.3).

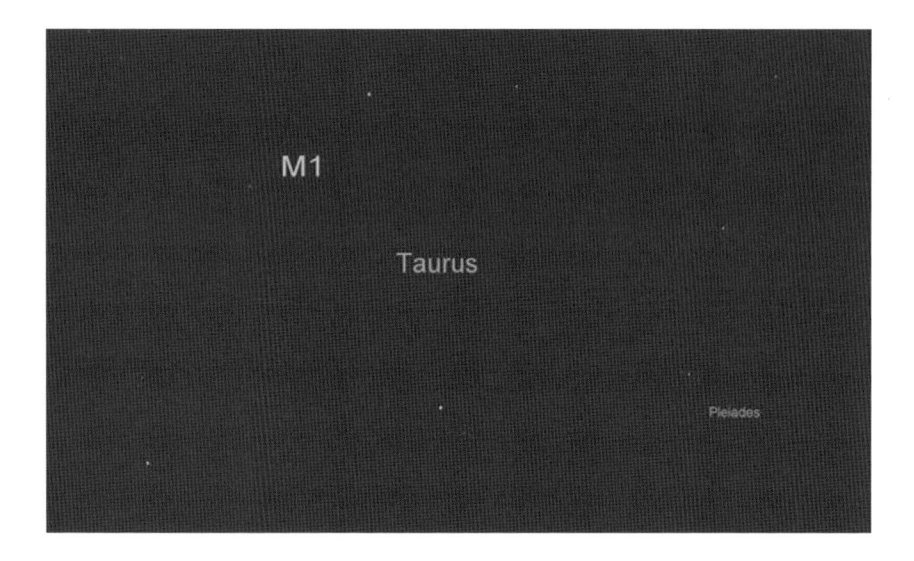

Fig. 3.2 How to find M1

Fig. 3.3 M1 modified

M2

This is a globular star cluster found in the northern reaches of Aquarius. It was discovered by Maraldi in 1746. It is very close to the celestial equator and far enough from the ecliptic that most solar system objects will not come near it. Through modest instruments, it doesn't look any different from an elliptical galaxy, with a dense core of about 20 arcseconds across or a little larger than Saturn (as seen from Earth). Indeed, it is hard to see its full extent of 16 arcminutes across (about half the apparent diameter of the Moon). It is about 37,500 light years away.

Due to extinction, the fainter nearby globular star cluster of M15 shows itself more readily to large binoculars on an average night.

Where to Find

M2 is hard to find directly, as it is in the faint constellation of Aquarius, whose stars are south of the celestial equator. The best reference point is Epsilon Pegasi (Elioth) and it is easier to find M15 which is fainter but more northerly and move almost due south to find M2. M15 actually appears brighter on most nights from England.

Fig. 3.4 M2 (Courtesy of Ray Grover, Manchester Astronomical Society)

From the southern hemisphere, it is easier to find M2 by moving south west of Elioth first (Fig. 3.5).

Object type	Globular star cluster
Declination	–0° 49 min
Right Ascension	21 h 33.5 min
Magnitude	6.5

What It Looks like

Through modest instruments, M2 looks like an elliptical galaxy and, in many cases, you will only see the core, especially on a moonlit night or one with thin cloud.

Charles Messier's Original Notes

11 September 1760
(RA: 21 h 21 m 08 s, 320d 17' 00"; Dec: –01d 47' 00"; Diameter: 0d 04')

Nebula without a star in the head of Aquarius, the centre is brilliant, and the surrounding light is round; she resembles the beautiful nebula located between the head and bow of Sagittarius and is very well seen with a telescope of 2 ft, placed upon the parallel with α Aquarii. M. Messier has reported this nebula on the chart of the comet observed in 1759. *Memoirs of the Academy for the year 1760, page 464.* M. Maraldi saw this nebula in 1746 by observing the comet which also appeared in that year.

How to View

With a large apparent size of 16 arcminutes, it is best not to use too high magnification, as 150× will fill the field of view with most commonly-used eyepieces. However, with apertures under 200 mm, it is better to use magnifications in the 30×to 50× range, or the extended parts of the cluster are not visible. Use of a light pollution reduction (LPR) filter definitely helps.

Photographic Details

Figure 3.4 shows how M2 appears in a 127 mm Maksutov, with some of the outer areas visible (Fig. 3.6).

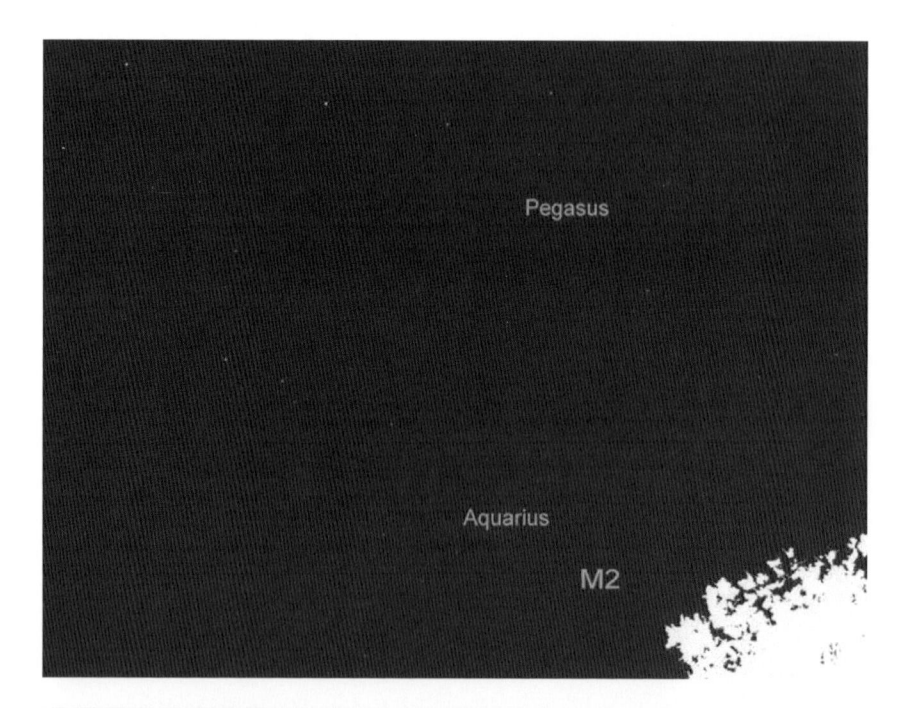

Fig. 3.5 How to find M2

Fig. 3.6 M2 Modified

M3

This is a globular star cluster found in Canes Venatici. It was discovered by Charles Messier in 1764. It is often overlooked for the better-known M13 in Hercules. It is apparently large at 18 arcminutes (although some photographs have suggested it could be larger). However, modest instruments only show its core and some of the surrounding area, suggesting a smaller size. It is actually a very large cluster and it is about 34,000 light years away, suggesting a diameter of about 180 light years and a mass approaching 250,000 suns. Its brightest stars are about magnitude 12.7, so can be seen in medium amateur instruments under good conditions.

It is said to be visible to the unaided eye but you are unlikely to see it from UK skies and due to its brightness being spread over a large area, it can be a challenging object for even good binoculars.

Although individual variable stars are not visible in a 127 mm telescope, advanced amateur and professional telescopes may detect some of them. 212 have been discovered to date, most of them RR Lyrae class stars, which have been used to estimate the cluster's distance. One of the great anomalies in cosmology is that some globular star clusters (notably this one) appear older than the age of the universe, according to some estimates. Could this mean that our current understanding of cosmology is wrong? Maybe not, but incomplete, certainly!

Fig. 3.7 M3 by Mike Deegan

Where to Find

Strictly speaking, M3 is within the boundaries of the small constellation Canes Venatici. However, it is best found using the same method of finding Arcturus from the handle of the Plough and stopping about half way (Fig. 3.8).

Object Type	Globular star cluster
Declination	28° 23 min
Right Ascension	13 h 42.2 min
Magnitude	6.2

What It Looks like

Through binoculars, M3 shows as a faint patch which is roughly circular, so really appears quite like an elliptical galaxy. Being somewhat fainter than M13, it isn't as easily resolved into stars as its neighbour and on most binocular viewings, it is only possible to make out the core. Although quite bright, a 127 mm Maksutov does not give a significantly better view than with binoculars.

Charles Messier's Original Notes

3 May 1764
(RA: 13 h 31 m 25 s, 202d 51' 19"; Dec: +29d 32' 57"; Diameter: 0d 03')

Nebula discovered between the Ox-herd [Boötes] and one of the hunting dogs of Hevelius [Canes Venatici]. It contains no star, the centre is brilliant and the light fades imperceptibly. It is round, and in a good sky you can see it with a telescope of 1 ft; it is brighter on a chart of the comet observed in 1779 (*Memoirs of the Academy,* the same year). Observed again on 29 March 1781, always very beautiful.

How to View

Like most globular star clusters, M3 is a tough object for most small instruments. It is highly likely that Charles Messier and astronomers of his era were only aware of it's central condensation. Although it is theoretically possible to see the outer regions in a 127 mm Maksutov, it would need a totally dark site and ideal conditions. It is more realistic to use apertures of 250 mm and above to see them.

Photographic Details

Figure 3.8 was taken using a 10" F5 Newtonian, Starlight Xpress HX916, guided with a 80ED refractor with a modifed Toucam. It consisted of 12 3-min exposure luminance frames, six further 80 s exposures luminance frames and ten 80 s exposures of RGB.

This photograph of M3 was taken using an SBIG ST2000XCM (cooled to 25° below ambient temperature) with an FLT110 APO Refractor and Astronomik LPR filter. 18 1-min subframes were stacked (Figs. 3.9 and 3.10).

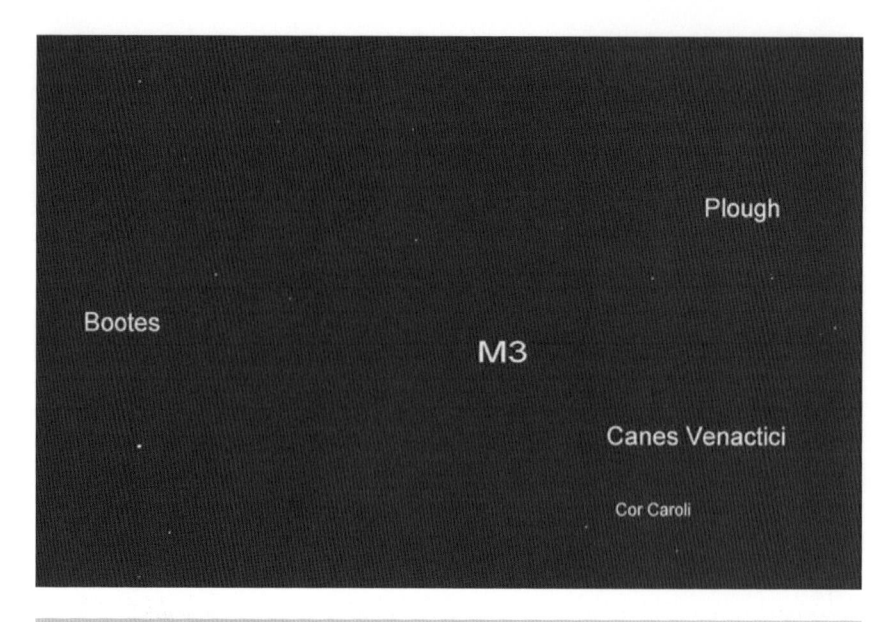

Fig. 3.8 How to find M3

Fig. 3.9 M3 modified

Fig. 3.10 M3 by Anthony Glover

M4

M4 is a globular star cluster near Antares in Scorpius. It was discovered by Philippe Loys de Chéseaux in 1746. Its angular size is 36 arcminutes, although small instruments should only detect the central condensation of about 14 arcminutes. It is about 7,200 light years away. Were M4 further north, it would be bigger and brighter than M13 but its low declination means that it never rises more than 13.5° above the horizon from the UK. It is easy to find (in theory) because it is close to Antares but extinction, light pollution and the background Milky Way make it a hard object from mid-northern latitudes, although it looks completely different when viewed from the southern hemisphere.

Being only 3° from the ecliptic, it means that solar system objects may pass close to it or (partially) occult it but the Moon and any of the brighter planets would "drown out" its light.

Where to Find

M4 is easy to find as it is very close to Antares, about a degree due west (Fig. 3.11).

Fig. 3.11 M4 (NASA/courtesy of nasaimages.org)

Object Type	Globular star cluster
Declination	−26° 32 min
Right Ascension	16 h 23.6 min
Magnitude	6.4

What It Looks like

Through binoculars, M4 appears like a brightening of the background Milky Way about half the size of the Moon. There is nothing to suggest it isn't an elliptical galaxy. Through a 127 mm Maksutov, it appears like a diffuse glow, not dissimilar from Omega Centauri.

Charles Messier's Original Notes

8 May 1764
(RA: 16 h 09 m 08 s, 242d 16′ 56″; Dec: −25d 55′ 40″; Diameter: 0d 02.5′)

Clusters of stars, very small; Which with a weak telescope appears in the form of a nebula; this star cluster is located near Antares and on its parallel. Observed by M. de la Caille, and reported in his catalogue. Reviewed 30 January and 22 March 1781.

How to View

Although M4 is quite bright, its large angular size means that it is not amenable to high magnification, especially from mid-northern latitudes. Small instruments do not have the light gathering power, especially as extinction is a problem when viewed from well north of the equator. A light pollution reduction (LPR) filter is a help but a larger instrument in the 200 mm class with low magnification is the best bet, although it is visible (albeit as a fuzzy patch) in my binoculars under good conditions.

Photographic Details

Due to the low declination as seen from the UK, it is difficult to spot individual stars through a 127 mm Maksutov, although the globular nature is obvious (Figs. 3.12 and 3.13).

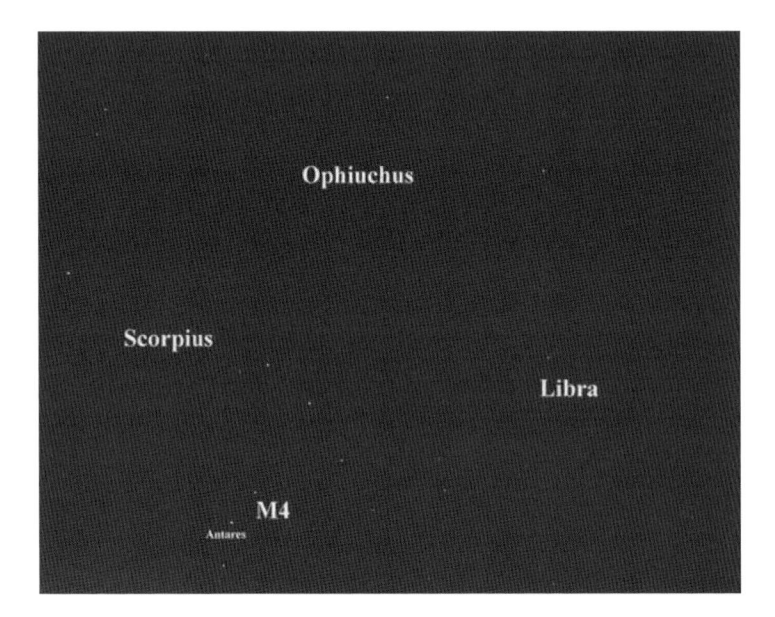

Fig. 3.12 How to find M4

Fig. 3.13 M4 modified

M5

M5, like the previous objects in the Messier Catalogue is a globular star cluster in Serpens. It was discovered by Gottfried Kirch and his wife, Maria in 1702. Although it has an angular size of 23 arcminutes, it is usually only the central condensation of about half of this that can be seen through modest instruments. At a distance of 24,500 light years, this corresponds to a diameter of 165 light years. In theory, it is an easy object, given its declination and magnitude but I've found that it doesn't always show well and there are few bright stars to act as guideposts.

Where to Find

M5 is found in a rather patchy area of sky but forms a rough equilateral triangle with Alpha and Mu in the faint constellation of Serpens Caput. It is close to the

Fig. 3.14 M5 (Courtesy of Kathleen Ollerenshaw, Manchester Astronomical Society)

border with Virgo and is near to 110 Virginis. Fortunately, it is visible in a reasonable finderscope (Fig. 3.14).

Object Type	Globular star cluster
Declination	2° 5 min
Right Ascension	15 h 18.6 min
Magnitude	6.1

What It Looks like

Although technically slightly brighter than nearby M13, it is further south and appears somewhat smaller. Only the central part can be seen in binoculars and it is tough for anything with an aperture of less than 50 mm. A 127 mm Maksutov armed with a wide angle eyepiece, focal reducer and light pollution reduction filter brings it to life somewhat and the central condensation and sparser outer regions becomes easy to make out but no stars are resolved.

Charles Messier's Original Notes

23 May 1764
(RA: 15 h 06 m 36 s, 226d 39' 04"; Dec: +02d 57' 16"; Diameter: 0d 03')

Beautiful nebula discovered between Libra and Serpens, near the star of the Serpent, of 6th magnitude or of the 5th, according to the Catalogue of Flamsteed: It does not contain any star, it is round, and can be seen very well in a good sky with an ordinary telescope of 1 ft. M. Messier has reported it on the chart of the Comet of 1763. *Memoirs of the Academy for the year 1774, page 40.* Reviewed on 5 September 1780, 30 January and 22 March 1781.

How to View

M5 is much more promising in larger instruments than I have. With modest apertures in the 70–150 mm range, a light pollution reduction (LPR) filter is definitely a help. The field of view should be kept to about 30 arcminutes, which allows moderate magnifications to be used with most instruments.

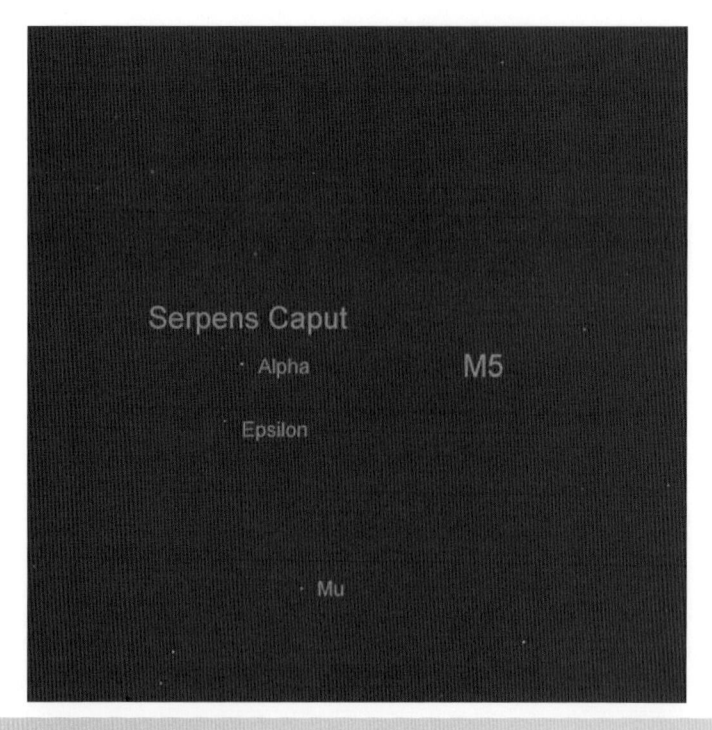

Fig. 3.15 How to find M5

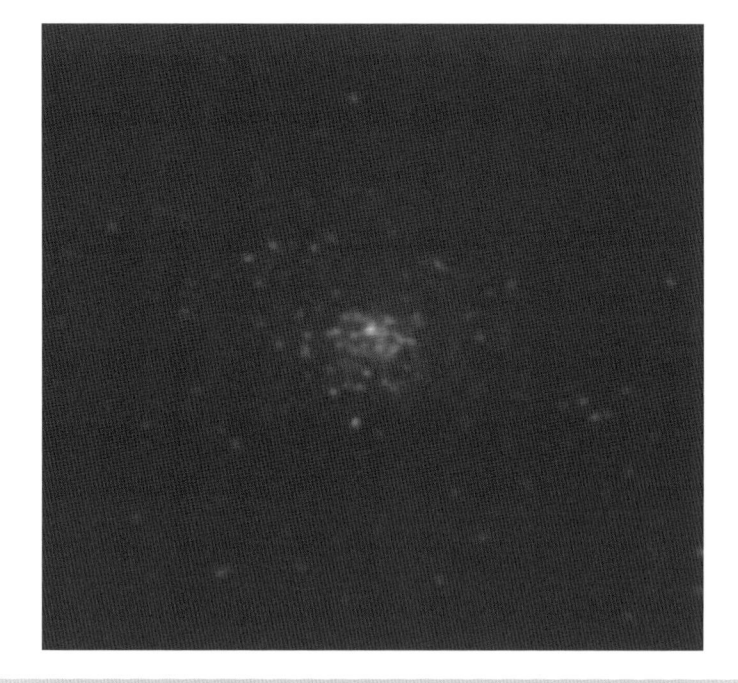

Fig. 3.16 M5 modified (Courtesy of Kathleen Ollerenshaw, Manchester Astronomical Society)

Photographic Details

Figures 3.15 and 3.16

M6

M6 in an open star cluster in Scorpius. It was discovered by Giovanni Battista Hodierna sometime before 1654. It is quite large, with an angular size of 54 arcminutes, although the central part is more like 25 arcminutes across and is 1,600 light years away. Unfortunately, it is too far south for reasonable viewing from the UK and rises barely 6°. Even from France and many parts of the United States, it is a difficult object. It is hardly surprising that my first view of it was from the southern hemisphere (Canberra, Australia) through 50 mm binoculars and the view was rather memorable.

Fig. 3.17 M6 and M7 (Courtesy of Ray Grover, Manchester Astronomical Society)

Where to Find

M6 is a tough object to find from southern England and Canada, although from the United States it can be found about 6° from the top of the scorpion's sting. From the more northern latitudes, it is best to find M8 first or M7. M7 is about 5° west of Epsilon Sagittari and about 10° south of M8 but slightly to the west. M6 is about 3° north west of M7. You may also find the "How to find M8" diagram useful, too. This shot (also used to find M7) was taken while on a business trip near Chicago and cannot be obtained from the UK (Fig. 3.17).

Object Type	Open star cluster
Declination	−32° 13 min
Right Ascension	17 h 40.1 min
Magnitude	4.2

What It Looks like

Discounting views from "Down Under", as viewed from England using 70 mm binoculars, M6 appears like a large open star cluster with lots of stars blinking in and out of view over an area somewhat smaller than the full moon. On a clear night, a few more stars come into view in a Maksutov. It is interesting to compare with M7. M7 is brighter but its area is larger, so the "surface brightness" appears about the same for both clusters.

Charles Messier's Original Notes

23 May 1764
(RA: 17 h 24 m 42 s, 261d 10′ 39″; Dec: −32d 10′ 34″; Diameter: 0d 15′)

Cluster of small stars between the bow of Sagittarius and the tail of Scorpius. To the naked eye this cluster seems to form a nebula with no stars, but with any instrument employed to examine it, one sees it as a cluster of small stars.

How to View

Apart from the obvious advice to travel south, going further than Paris or Prague would be "cheating" in the Messier sense. A clear horizon and large aperture helps but you really need a full degree field of view to see the whole cluster. This means that a short focal length or focal reducer is necessary. As some of the cluster members are quite faint, a small telescope is better than binoculars.

Photographic Details

Figures 3.18 and 3.19

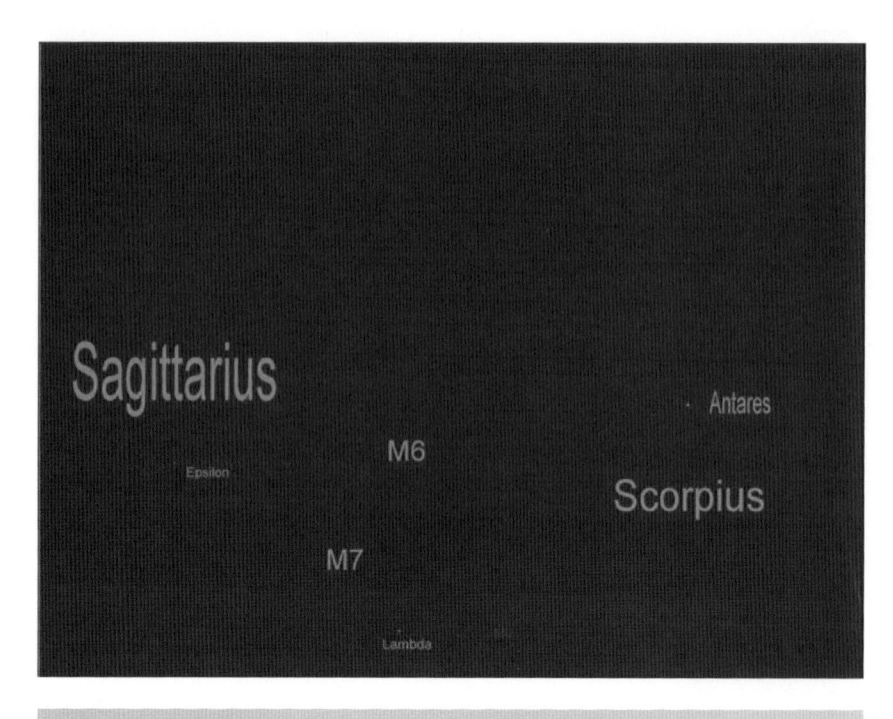

Fig. 3.18 How to find M6

Fig. 3.19 M6 modified

M7

This is an open star cluster in Scorpius's tail. It was known to Ptolemy in 130 AD and is sometimes known as Ptolemy's Cluster. It has a large apparent diameter of 1.3° but its low southerly declination renders it all but invisible from the UK and difficult from the United States. Like M6, my first view of it was from Canberra, Australia with 50 mm binoculars. Although brighter than M6, it should appear about the same brightness, as its light is spread out over a larger area. It is about 800 light years away.

Where to Find

M7 is easy to find from southern latitudes, being about 4° east and 2° north of the scorpion's tail. It is also large and bright and not easy to miss. From more northern latitudes, it is found by going about 5° west of Epsilon Saggittari or 10° south of M8 and slightly west. You may find the "How to find M8" diagram useful, too (Fig. 3.20).

Fig. 3.20 M7 modified (NASA/courtesy of nasaimages.org)

Object Type	Open star cluster
Declination	−34° 49 min
Right Ascension	17 h 53.9 min
Magnitude	3.3

What It Looks like

From England using 70 mm binoculars, M7 appears faint but large, nowhere near the spectacular sight it does from a site where it attains a greater elevation. The stars can just about be made out.

In a 127 mm Maksutov, a few more stars become visible but, honestly, without knowing that it happens to be a star cluster, one would be forgiven for thinking it was just a rich part of the Milky Way.

Charles Messier's Original Notes

23 May 1764
(RA: 17 h 38 m 02 s, 264d 30′ 24″; Dec: −34d 40′ 34″; Diameter: 0d 30′)

Star cluster greater than the preceding; this cluster appears to the naked eye as a single nebula. It is not far from the preceding [M6], lying as it is between the bow of Sagittarius and the tail of Scorpius.

How to View

With the difficulty in maneuvering even a small telescope to a position where M7 can be viewed, hand-held binoculars make the task so much easier. A few more stars become visible in larger apertures but the binocular view is more aesthetically pleasing.

Photographic Details

M7 can be seen in the same photograph as M6. See Fig. 3.18 for details (Fig. 3.21).

M8

Fig. 3.21 How to find M7

M8

M8 is a superb example of a star-forming region, like the Great Orion Nebula (M42) and is also known as the Lagoon Nebula. It was discovered by Giovanni Battista Hodierna sometime before 1654. Unfortunately, it is never seen at its best from most northern latitudes, so loses a lot of its light through extinction. On very clear nights, it can be equally difficult, as its brightness is spread out over a large area (90×40 arcminutes) and there's the glow of the background Milky Way. It is about 5,200 light years away. It may also be confused with the open star cluster NGC6530, which would have almost certainly made it into the Messier Catalogue it its own right, had it not been for its southerly declination. In fact, NGC6530 is physically associated with M8. However, larger instruments separate the pair.

Dark dust lanes make the nebula to appear broken in half and some smaller dust lanes are visible in larger amateur instruments.

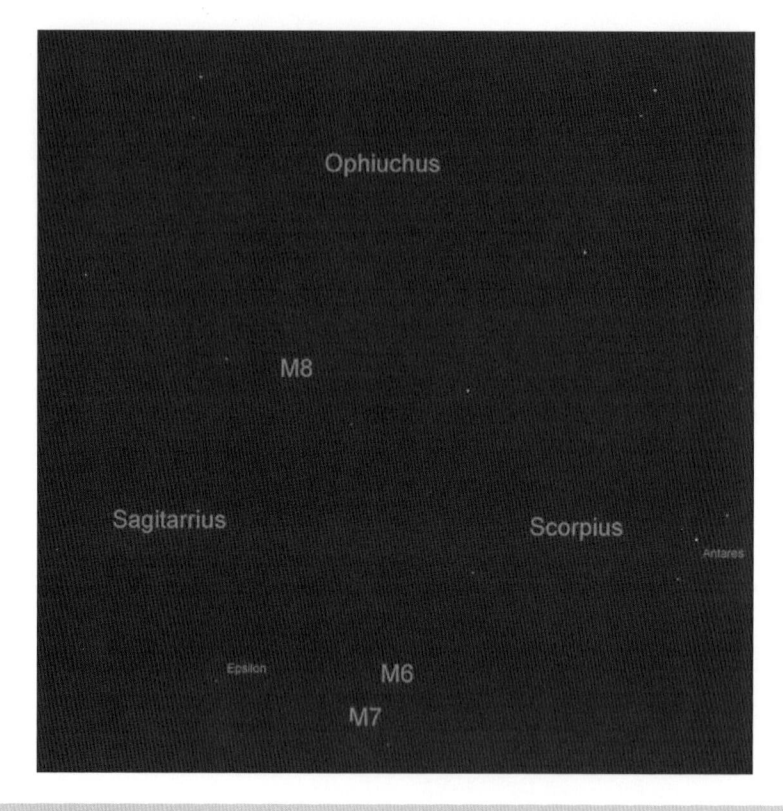

Fig. 3.22 How to find M8

Where to Find

M8 lies in a rather sparse region of sky and there are no bright stars near. However, there's some 6th and 7th magnitude stars that appear almost in a line and the cigar-shaped glow of M8 is quite easy to see in binoculars or even a finderscope (Fig. 3.22).

Object Type	Nebula
Declination	−24° 23 min
Right Ascension	18 h 3.8 min
Magnitude	6.0

What It Looks like

M8 looks surprisingly prominent in 70 mm binoculars. It appears as a bright extended patch of light in a generally rich part of the sky. It looks like a large "cigar" shape, possibly suggesting an elliptical galaxy or edge-on spiral galaxy.

On a clear night in a 127 mm Maksutov, it is simply spectacular. Not only is the nebula visible but myriads of stars, some of which belong to NGC6530 and others that are foreground/background stars. The view is reminscent somewhat of the pictures of the Rosette Nebula on websites and in magazines. If you use too high a magnification and the sky is a bit twilit, it can appear faint and diffuse. However, it is one of the more spectacular objects and can become addictive and keep drawing you back to it.

Charles Messier's Original Notes

23 May 1764
(RA: 17 h 49 m 58 s, 267d 29' 30"; Dec: −24d 21' 10"; Diameter: 0d 30')

Star cluster that appears in the form of a nebula when looking with an ordinary telescope of 3 ft; but with an excellent instrument there are a lot of small stars; near to this cluster is a bright star surrounded by a faint glow and is the 9th star of Sagittarius [9 Sagittarii], 7th magnitude, according to Flamsteed: this cluster appears as an elongated shape that extends from north-east to south-west, between the bow of Sagittarius and the right foot of Ophiuchus.

How to View

It is best to see M8 in its entirety, which means a large field of view. This limits the magnification to about 20–30x for most telescopes. It is bright enough to be enjoyed in binoculars. Large instruments show more detail, such as the dust lanes, although the effect of the whole object becomes lost.

Photographic Details

In a 127 mm Maksutov, some of the fainter areas of M8 appear as dark background, whilst the brighter regions give a wispy appearance (Figs. 3.23 and 3.24).

Fig. 3.23 M8 (Courtesy of Ray Grover, Manchester Astronomical Society)

Fig. 3.24 M8 modified (NASA/courtesy of nasaimages.org)

M9

Like many of the globular star clusters concentrated near the centre of the Milky Way (as seen from Earth), M9 never gets particularly high in the sky, although it can attain an elevation of 20° from southern England. It was discovered by Charles Messier in 1764. Its apparent size is a large 12 arcminutes, although small instruments can only make out its core of about a quarter of that size. It is about 25,800 light years away. Instruments in the 200 mm class can resolve some of the stars (the brightest of which are comparable to Pluto in magnitude) and also the outlying regions. Due to dark material in the plane of the Milky Way, it is actually about 10 times fainter than it would be if the view was unobstructed, so would be about magnitude 5.2.

Where to Find

M9 is quite a hard object to find. The nearest bright star is Eta Ophiuchii. M9 is about 3° south and a degree east (Fig. 3.25).

Fig. 3.25 M9 (NASA/courtesy of nasaimages.org)

Object Type	Globular star cluster
Declination	−18° 31 min
Right Ascension	17 h 19.2 min
Magnitude	8.1

What It Looks like

Through small instruments, M9 is more one of the items to cross off a list of things to see than one of the more memorable members of the Messier catalogue. In 70 mm binoculars it was almost invisible even on a particularly clear night and on an averagely "clear" night would probably be missed. A 127 mm Maksutov reveals a larger, diffuse object and there's little to suggest it is a globular star cluster, as opposed to a galaxy or distant nebula.

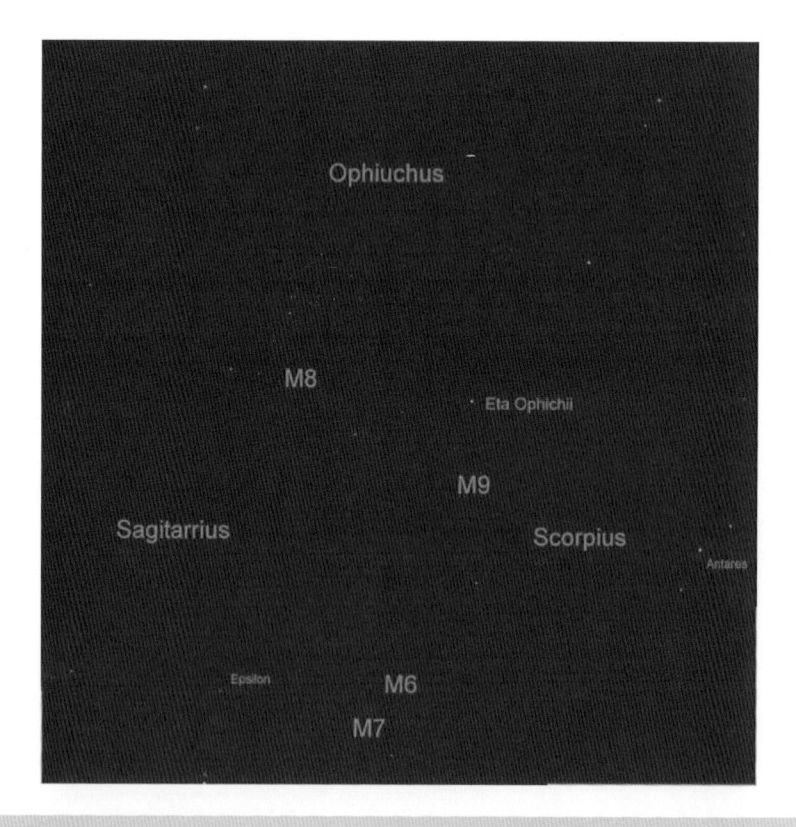

Fig. 3.26 How to find M9

M9

Charles Messier's Original Notes

28 May 1764
(RA: 17 h 05 m 22 s, 256d 20' 36"; Dec: −18d 13' 26"; Diameter: 0d 03')

Nebula without a star, in the right leg of Ophiuchus, she is round and her light is faint. Observed again on 22 March 1781.

How to View

M9 is not really a suitable object for modest instruments. It really needs a larger aperture telescope of at least 200 mm to do it justice. Quite high magnification of 100x can be used to resolve some of the individual stars.

Photographic Details

Figures 3.25 and 3.27

Fig. 3.27 M9 (Modified NASA/courtesy of nasaimages.org)

M10

M10 is one of the brighter globular star clusters in Ophiuchus and most of its brightness is concentrated into 8 or 9 arcminutes, although its true size is nearer to 15 arcminutes. It was discovered by Charles Messier in 1764. It is 14,300 light years away. Being near the celestial equator, it can attain an elevation in excess of 30° and can be as high as 50° from the southern United States. This certainly brings it within range of many binoculars. From my suburban location, I have only seen it a handful of times in my 15×70 binoculars.

Where to Find

M10 lies in Ophiuchus and is about half a degree west of the 5th magnitude star 30 Ophiuchi. Fortunately, it is quite bright and visible in a finderscope on a good night (Fig. 3.28).

Fig. 3.28 M10 (NASA/courtesy of nasaimages.org)

M10

Object Type	Globular star cluster
Declination	−4° 6 min
Right Ascension	16 h 57.1 min
Magnitude	6.6

What It Looks like

M10 and M12 can be seen in the same binocular field of view and appear quite clearly as roughly circular with a central condensation of stars. They appear as twins and a 127 mm Maksutov shows the effect more clearly but suggests a slightly larger angular size than binoculars.

Charles Messier's Original Notes

29 May 1764
(RA: 16 h 44 m 48 s, 251d 12′ 06″; Dec: −03d 42′ 18″; Diameter: 0d 04′)

Nebula without a star, in the belt of Ophiuchus, near the thirtieth star of this constellation, according to Flamsteed, and of 6th magnitude. This nebula is beautiful and round; it can be difficult to see with a regular telescope of 3 ft. M. Messier has reported it on the 2nd chart of the comet of 1769. *Memoirs of the Academy for the year 1775.* Plate IX. Observed again on on 6 March 1781.

How to View

Although M10 can be seen in modest instruments, it is best seen in large apertures with a 20 arcminute or greater field of view. This allows some of the outer stars to be resolved. A magnification of around 100× can be used, depending on the focal ratio and focal length of the telescope.

Photographic Details

Figures 3.28 and 3.30

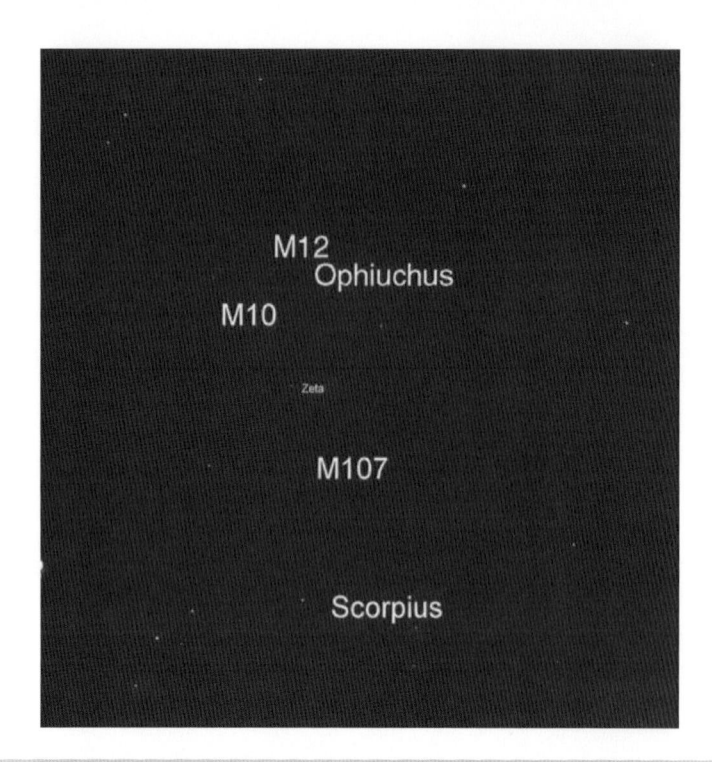

Fig. 3.29 How to find M10

Fig. 3.30 M10 (Modified NASA/courtesy of nasaimages.org)

M11

M11 is an open star cluster in the southern constellation of Scutum. Its nickname is the Wild Duck cluster because the shape of its brighter components suggests the wings of a flying duck. When you see it for the first time, it is quite easy to think its part of the Milky Way, although it is slightly closer to us than the background stars. It measures 14 arcminutes across, at its widest part but some of the outlying fainter stars can be missed under poor conditions or with small aperture instruments. It is 6,000 light years away. It was discovered by the German astronomer Gottfried Kirch in 1681.

Where to Find

Although M11 is in Scutum, it is a lot easier to find by moving south from Altair about 8°. It is very close to the border with Aquila.

Object Type	Open star cluster
Declination	−06° 16 min
Right Ascension	18h 51.1 min
Magnitude	6.3

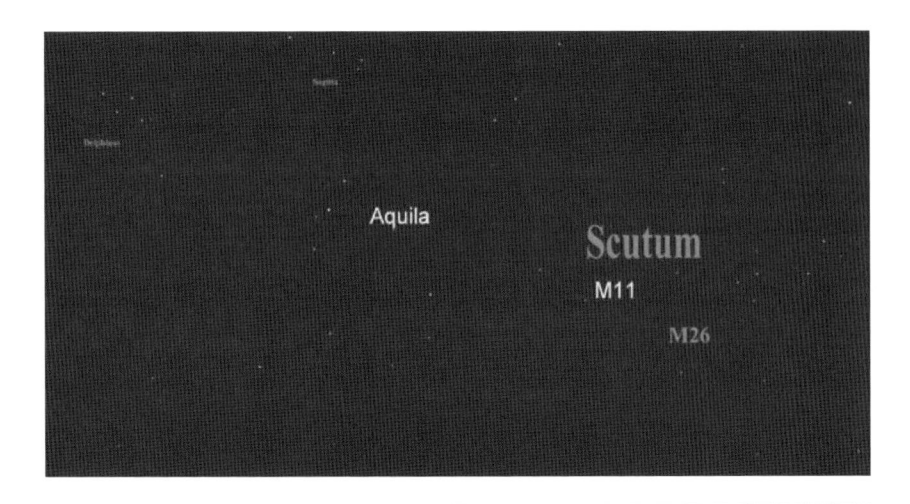

Fig. 3.31 How to find M11

What It Looks like

Through very small aperture instruments, it does fit its nickname but the wild duck effect really needs a larger instrument to bring out the best in it. As you move up the apertures, the outlying fainter stars come into view. Sometimes, it stands out better on less than perfect nights, as the background Milky Way is very rich in this part of the sky. The addition of a light pollution reduction filter really brings M11 to life from suburban skies.

Charles Messier's Original Notes

30 May 1764
(RA: 18 h 30 m 23 s, 279d 35' 43"; Dec: −06d 31' 01"; Diameter: 0d 04')

Clusters of many small stars, near the star K of Antinous [a defunct constellation that is now part of Aquarius], that can be seen with a good instrument; with an ordinary telescope of 3 ft it looks like a comet: The cluster is mixed with a faint light; in this cluster there is a star of 8th magnitude. M. Kirch observed it in 1681. *Philof Transact. No. 347, page 390*. It is reported in the *Great English Atlas*.

How to View

The easiest way to view M11 is by using large aperture binoculars or small, short focal length refractors. It is not easy to find, even with a comparatively large field of view through my Maksutov-Cassegrain, indeed, as well as having problems finding the object, it was also difficult to make out against the Milky Way background, which my Maksutov-Cassegrain just brightens. Although it can be detected in instruments as small as 21 mm binoculars, the best view is through my 80 mm short tube refractor, armed with a 32 mm eyepiece to yield a 4° field of view, although large, tripod-mounted binoculars would probably be equally effective.

Photographic Details

This is a good representation of how M11 looks through a 127 mm Maksutov, so needs no modification (Fig. 3.32).

Fig. 3.32 M11 (Courtesy of Joe Billington, Manchester Astronomical Society)

M12

M12 is very close to M10 and, for the less experienced viewer, it can even be confused with it. It is slightly fainter than M10 and an arcminute larger, spreading its light over a larger area. It is 16,000 light years away. It is also much less concentrated towards the center as M10 and most other globular star clusters, so for a while, there was some doubt as to whether it was a true globular or some sort of "intermediate type" between globular and open. As it is slightly nearer the celestial equator than M10, when near the horizon may appear brighter, as it suffers less. Indeed, it can attain an altitude of 37° from southern England. It was discovered by Charles Messier in 1764.

Where to Find

M12 lies in Ophiuchus and is about 3° north west of M10, which is slightly easier to find. Fortunately, it is quite bright and visible in a finderscope on a good night (Fig. 3.33).

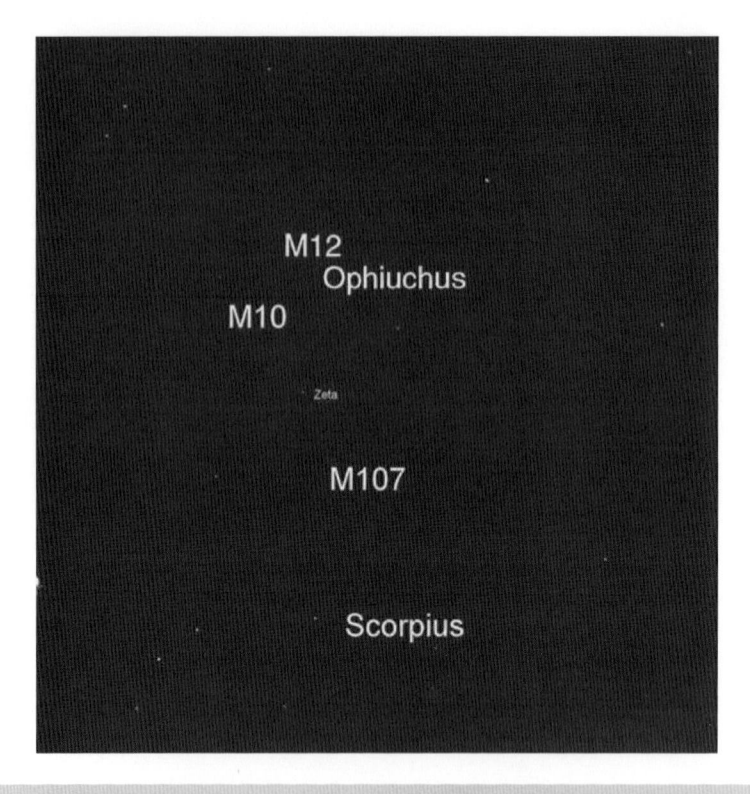

Fig. 3.33 How to find M12

Object Type	Globular star cluster
Declination	−1° 57 min
Right Ascension	16 h 47.2 min
Magnitude	7.1

What It Looks like

M10 and M12 can be seen in the same binocular field of view and appear quite clearly as roughly circular with a central condensation of stars. They appear as twins and a 127 mm Maksutov shows the effect more clearly but suggests a slightly larger angular size than binoculars.

Charles Messier's Original Notes

30 May 1764
(RA: 16 h 34 m 53 s, 248d 43′ 10″; Dec: −02d 30′ 28″; Diameter: 0d 03′)

M12

Nebula discovered in the serpent [Serpens], between the arm and the left side of Ophiuchus: This nebula contains no star, it is round and her light is faint; near this nebula is a star of 9th magnitude. M. Messier reported it on the second chart of the comet observed in 1769. Memoirs of the Academy for the year **1775, plate** *IX*. Observed again on 6 March 1781.

How to View

Although M12 can be seen in modest instruments, it is best seen in large with a 20 arcminute or greater field of view. This allows some of the outer stars to be resolved. A magnification of around 100x can be used, depending on the focal ratio and focal length of the telescope.

Photographic Details

Figures 3.34 and 3.35

Fig. 3.34 M12 (NASA/courtesy of nasaimages.org)

Fig. 3.35 M12 (Modified NASA/courtesy of nasaimages.org)

M13

M13 is sometimes known as the "Great Globular Cluster in Hercules", distinguishing it from the lesser known M92. It is the brightest globular star cluster north of the celestial equator and its high northern declination makes it circumpolar from northern parts of the United Kingdom and Canada, although extinction would make it invisible when it is within a few degrees of the horizon. It was discovered by Edmund Halley (of comet fame). Outlying members of this cluster give it an apparent size of 20 arcminutes, although its central 13 arcminutes is more readily visible. It is about 25,100 light years away. It has certainly been seen without optical aid by Halley himself and several people since, although it is just about impossible from modern light-polluted skies. Nevertheless, it is the easiest globular star cluster for inhabitants of the northern hemisphere, although it is much fainter than the brilliant Omega Centauri, which is visible from southern locations, such as Brazil.

In addition to being a popular target for amateur astronomers, it has also been well studied by researchers. Estimates of its age by examining its stellar population make it older than the generally accepted age of the universe of 13,700 million years. The most conservative of these estimates place it at 14,000 million years old.

Now M13 has a personal anecdote attached. During the early part of my astronomical renaissance, it was one of my "must see" objects. Unfortunately, it was near the top of my "couldn't see" objects. With a magnitude of 5.7 and a high northerly declination, it should have been easy enough, except it wasn't. Whilst I now know the value of a decent pair of binoculars, at the time, I had 20x50 binoculars of dubious quality. Within a week of the arrival of my new 15x70 binoculars, I had potted it! I wasn't immediately impressed but, on a clear night, I could see the core and the surrounding area and seen my first globular star cluster. Since then, I've seen a few more, including the very impressive Omega Centauri from Brazil. Were it not for its extreme southerly declination, it would certainly be in the Messier Catalogue.

Where to Find

M13 is found about a third of the way down the western side of the "keystone" asterism in Hercules. From suburban skies, the full constellation of Hercules is difficult to make out but its position between Lyra and Boötes makes it easier to find. If you can't see the keystone, then maybe conditions will not be good enough to see M13 (Fig. 3.36).

Object Type	Globular star cluster
Declination	36° 28 min
Right Ascension	16 h 41.7 min
Magnitude	5.7

Fig. 3.36 M13 by Anthony Glover

What It Looks like

Through small instruments, M13 appears as a fuzzy patch and may even be missed completely from suburban skies. 70 mm binoculars on a clear night start to resolve its nucleus, possibly suggesting that it is an elliptical galaxy. A 127 mm Maksutov starts to resolve individual stars on the outer fringes, whilst larger instruments can resolve details almost to the core.

A revisit with the Maksutov, armed with a light pollution reduction filter really brought it alive. There was no question that it was a globular star cluster and the appearance of stars on the outer fringes blinking into and out of existence was a remarkable sight.

Charles Messier's Original

1 June 1764
(RA: 16 h 33 m 15 s, 248d 18' 48"; Dec: +36d 54' 44"; Diameter: 0d 06')

Nebula without a star, discovered in the belt of Hercules; It is round and brilliant, the center being brighter than the edges as seen with a telescope of one foot; and is near two stars, one and the other of eighth magnitude, one above and another below the nebula was determined by comparing it to ε Herculis. M. Messier has reported it on the chart on the Comet of 1779, included in the *Memoirs of the Academy of the year 1784*. Seen by Halley in 1714. Observed again on 5 and 30 January 1781. It is reported in the *English Celestial Atlas*.

How to View

M13 is one of the larger and brighter Messier objects but presents its own challenges. Small instruments do not resolve individual stars, whilst larger instruments do not allow you to get the object into the field of view, although close-ups of various parts are worth enjoying for their own sake. For larger apertures, a long focal length eyepiece and/or focal reducer will allow you to see the entire object.

Photographic Details

Figure 3.37 shows M13. It was taken using a 80 mm/F6 refractor and SXV H9 Starlight Xpress camera cooled to 25° below ambient temperature. Six images, each of 1 min duration (for each of Luminosity, Red, Green and Blue) were taken and stacked.

Figure 3.38 shows M13 as seen through a 127 mm Maksutov.

Fig. 3.37 How to find M13

Fig. 3.38 M13 modified

M14

Although M14 is a globular star cluster in the vicinity of the better known M10 and M12, it is a somewhat different object, being more ellipsoidal in shape than the more usual spherical shape of most globular star clusters. It was discovered by Charles Messier in 1764. It is only just south of the celestial equator, so can attain an altitude of over 30° from the southern UK or Canada. It is smaller in size than its neighbors. It is about 30,300 light years away. Although it is 11 arc-minutes along its greater axis, the central 3 arcminutes is about all you can realistically expect in small to medium aperture instruments. It has been described as looking like an elliptical galaxy and, honestly, I think it's a fair assessment.

Where to Find

M14 is in a rather sparse patch of sky and is a full magnitude fainter than M10 and M12. It is about 8° east of M10 and about 15° south of Alpha Ophiuchii (Fig. 3.39).

Fig. 3.39 M14 (NASA/courtesy of nasaimages.org)

M14

Object Type	Globular star cluster
Declination	–3° 15 min
Right Ascension	17 h 37.6 min
Magnitude	7.8

What It Looks like

Compared to the nearby M12 and M10, M14 is a bit of a disappointment. A faint patch registered in 15×70 binoculars on a less than perfect night. In a 127 mm Maksutov, it takes on a rather wispy appearance, like a very faint Omega Centauri. It is not possible to resolve individual stars.

Charles Messier's Original Notes

1 June 1764
(RA: 17 h 25 m 14 s, 261d 18' 29"; Dec: –03d 05' 45"; Diameter: 0d 07')

Nebula without a star, discovered in the drapery that passes over the right arm of Ophiuchus, & placed on the [same] parallel [as] ζ Serpentis: this nebula is not large, with a faint light, we can see, however, with a telescope of three & a half feet; it is round, near it is a small star of the 9th magnitude; its position was determined by comparing it with γ Ophiuchi, & M. Messier has reported its position on the chart of the Comet of 1769. *Memoirs of the Academy of the year 1775, plate IX.* Reviewed 22 March 1781.

How to View

Like many globular star clusters of similar brightness, M14 really needs a larger aperture to do it justice.

Photographic Details

Figures 3.39 and 3.41

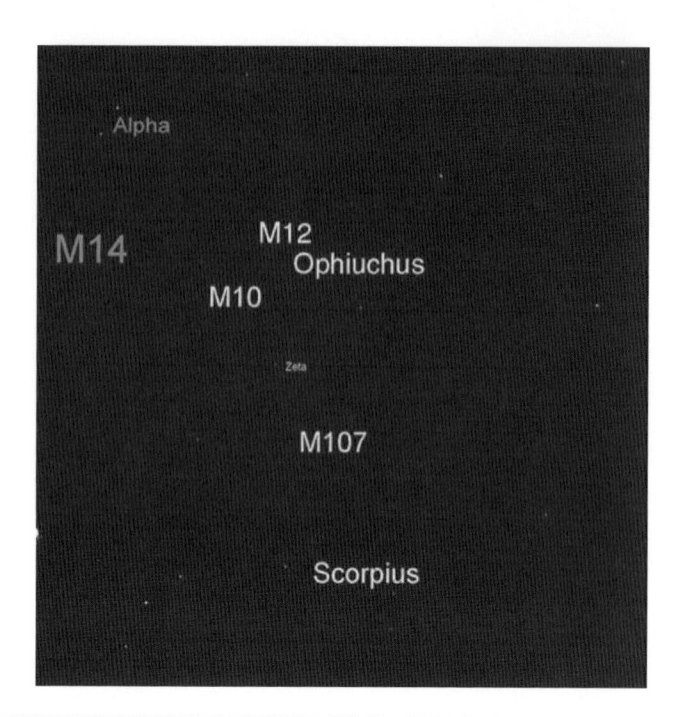

Fig. 3.40 How to find M14

Fig. 3.41 M14 (Modified NASA/courtesy of nasaimages.org)

M15

M15 is a globular star cluster on the border of Pegasus. It has an apparent diameter of 18 arcminutes, although its central 7 arcminutes is more readily visible to amateur instruments. Half of its mass is concentrated into a central area of about an arcminute. This seems to suggest either an extreme stellar density at the core, or even the existence of a medium-sized black hole. It is about 33,600 light years away. As it is in the same part of the sky as M2, comparisons are almost inevitable. From England, M15 is more readily visible than M2, despite being fainter. The higher northerly declination of M15 means it can attain altitude of 50° from southern England and most of Canada. From the southern United States, it can attain an altitude of more than 70°.

It is borderline visible to the unaided eye under favorable conditions, so it may be surprising that it wasn't discovered until 1746 by Jean-Dominique Maraldi. Its true nature as a globular star cluster was ascertained by William Herschel. Indeed, through most amateur instruments, there is nothing to distinguish it from an elliptical galaxy, unless some of the outer stars are visible.

Where to Find

The brightest star at the right of the picture is Epsilon Pegasi. Move 3.5° west and 2.25° north and you find it. It is about 13° north of M2 and just 3 min of right ascension to the west. If you center M2 in your finderscope or binoculars, a sweep of 13° to the north will bring it into your field of view. From the northern hemisphere, It is often easier to find M15 before finding M2 (Fig. 3.42).

Fig. 3.42 M15 by Anthony Glover

Object type	Globular star cluster
Declination	12° 10 min
Right Ascension	21 h 30.0 min
Magnitude	6.2

What It Looks like

Through small binoculars, M15 appears as a fuzzy patch on a good night. On a difficult night, it isn't necessarily visible even in 15×70 binoculars. It is only when you reach about 100 mm aperture, you can resolve some of the brighter outer stars and can say that it's a globular star cluster, rather than an elliptical galaxy.

Charles Messier's Original Notes

3 June 1764
(RA: 21 h 18 m 41 s, 319d 40' 19"; Dec: +10d 40' 03"; Diameter: 0d 03')

Nebula without a star, between the head of Pegasus and the "little horse" [Equuleus]; it is round, the centre is brilliant, its position determined by comparing it to δ Equulei. M. Maraldi, in *Memoirs of the Academy for the year 1746*, speaks of this nebula: "I saw it, between the star ε Pegasi & β Equulei, a nebulous star quite clear, which is composed of several stars; its right ascension is 319d 27' 6", & its northern declination is 11d 2' 22'''".

How to View

M15 definitely needs a cloudless night before even attempting an observation. Any cloud will render it invisible to binoculars and medium aperture telescopes will only show it as a roughly circular misty patch. To identify it as a globular star cluster, you need a reasonable magnification of 100× or higher to resolve the gaps between the outer stars and the gaps between the outer stars and the core. This is just about the limit of my 127 mm Maksutov-Cassegrain. If you have access to a larger aperture instrument, the view will improve.

Photographic Details

M15 was taken using a 110 mm/F7 refractor and SBIG ST2000XCM camera cooled to −30°. Twenty images, each of 1 min duration were taken and stacked and processed using *Nebulosity* (Stark Software) (Fig. 3.42).

Figure 3.44 is a more realistic visual view through a small telescope, where only some of the outer stars are resolved and the core becomes more fuzzy.

Fig. 3.43 How to find M15

Fig. 3.44 Modified M15

M16

M16 actually consists of two objects, an open star cluster (NGC 6611) and nebula (IC 4703) in Serpens Cauda. NGC 6611 was discovered by Philippe Loys de Chéseaux around 1745 and IC 4703 was discovered by Charles in 1764. The two objects are physically associated. IC 4703 is known as the Eagle Nebula and was made famous by the Hubble Space Telescope with publication and wide circulation of the "Pillars of Creation" photograph. This shows a close-up where star formation is actively taking place. Although no such detail is visible in amateur instruments, the pillars can be resolved in instruments as small as 300 mm on a good night. NGC 6611 has an angular size of only 7 arcminutes (about a quarter of the Moon's diameter). As its brightest stars are about 8th magnitude, they should be readily visible to modest instruments, although the southerly declination does not help matters from the UK and Canada.

IC 4703 is much larger, about the same size as the Moon. It is very similar in nature to the Great Orion Nebula (M42) but much further away (6–7,000 light years) and about 15° further south.

Where to Find

M16 and M17 are quite close. Whilst there's no obvious signposts, they appear brighter than their published magnitudes would suggest. Using binoculars, a south west sweep from Altair through M11 will find either or both of them if you strike lucky and get them in the same field of view. M16 is about 2.5° north of M17 (Fig. 3.45).

Object Type	Open star cluster and nebula
Declination	−13° 47 min
Right Ascension	18 h 18.8 min
Magnitude	6.4

What It Looks like

If you are familiar with the Pillars of Creation photos and those produced by advanced amateurs with good equipment, M16 will appear disappointing. Nevertheless, if you like M42 and M8, you'll like this. It appears as a fuzzy patch in binoculars but a 127 mm Maksutov shows it more clearly and, from an aesthetic point of view, getting the cluster and background stars into view makes it all the more enjoyable.

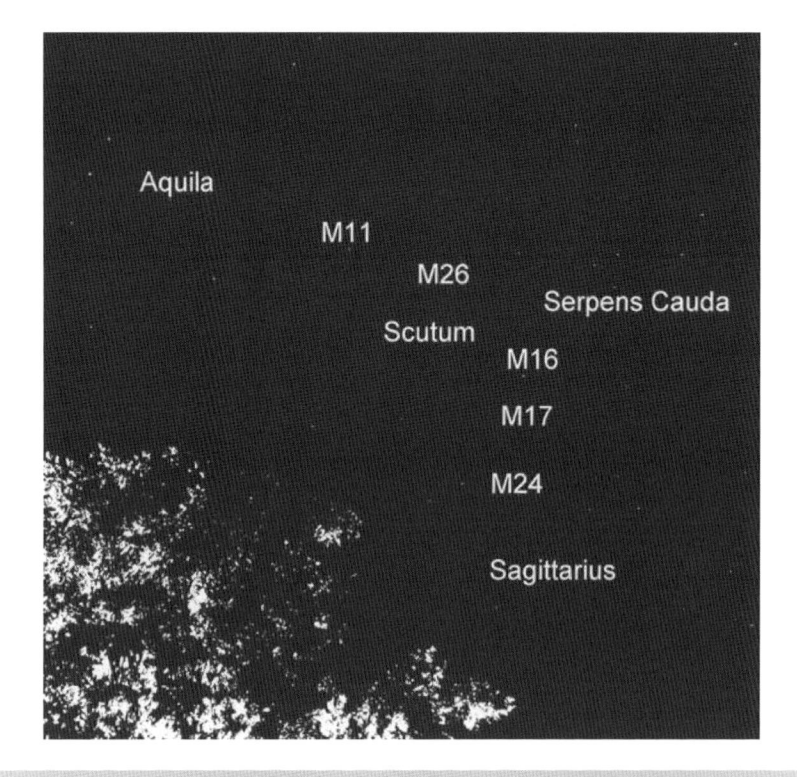

Fig. 3.45 How to find M16

Charles Messier's Original Notes

3 June 1764
(RA: 18 h 05 m 00 s, 271d 15′ 03″; Dec: −13d 51′ 44″; Diameter: 0d 08′)

Cluster of small stars, blended with a faint light near the snake's tail [Serpens], a short distance to the parallel of the ζ star of this constellation; faint with a telescope this cluster appears as a nebula.

How to View

Whilst there is little doubt that a large aperture helps, especially when it comes to seeing anything like the view as shown in photographs, its appearance in smaller instruments is rather better than that of many globular clusters and fainter galaxies. It is certainly worth visiting with anything in excess of 50 mm aperture and it will also take magnifications up to 100×.

Photographic Details

Figure 3.46 was taken by Anthony Glover using a 110 mm F/7 refractor, with an SXV H9 camera. Six 5 min exposures were taken using a hydrogen alpha filter.

Figure 3.47 shows a more realistic representation of M16. It isn't as clear as Anthony's picture but the main structure can be seen clearly.

Fig. 3.46 M16 by Anthony Glover

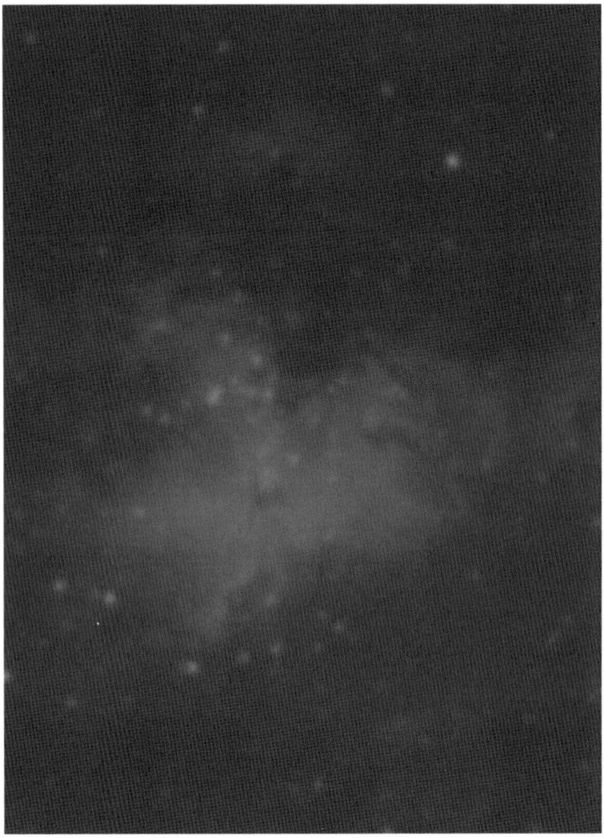

Fig. 3.47 M16 modified

M17

M17 has several nicknames, possibly the most commonly used one being the Swan Nebula. It is also known as the Omega, Horseshoe or Lobster Nebula. It is brighter and smaller than the Eagle Nebula (M16/IC 4307) but in a similar region of sky. Like the Eagle Nebula, its southern declination makes it a difficult object from the UK and Canada, except on clear nights. From southern parts of the United States, it can reach a more healthy 40° above the horizon. It was discovered by Philippe Loys de Chéseaux around 1745 but Charles Messier discovered it independently in 1764. It may have been known about in earlier times, as it has been recorded as being visible to the unaided eye under favourable locations from southern latitudes.

It is slightly closer than the Eagle Nebula, at a distance of about 5,000 light years and is in the same spiral arm of the Milky Way. It is therefore possible that the two objects can be physically associated and part of a large interstellar star-forming region. Its actual size is much larger than the more well-known Great Orion Nebula (M42), which happens to be nearly four times closer and easier to see from the northern hemisphere.

Where to Find

M16 and M17 are quite close. Whilst there's no obvious signposts, they appear brighter than their published magnitudes would suggest. Using binoculars, a south west sweep from Altair through M11 will find either or both of them if you strike lucky and get them in the same field of view. M17 is about 2.5° south of M16 (Fig. 3.48).

Fig. 3.48 M17 (Courtesy of Ray Grover, Manchester Astronomical Society)

M17

Object Type	Nebula
Declination	−16° 11 min
Right Ascension	18 h 20.8 min
Magnitude	6.0

What It Looks like

Through 70 mm binoculars, M17 looks like a fuzzy patch. With its proximity to M16 in terms of both the physical sky and catalogue, it is inevitable that you will compare the two. M17 is a bit brighter and smaller, so the nebula effect is better than M16, so appears better in binoculars. A 127 mm Maksutov brings out the nebula more but it doesn't have the same rich stellar foreground and background of M16.

To sum up they are both items to visit and revisit, rather than cross of the list of things to see.

Charles Messier's Original Notes

3 June 1764
(RA: 18 h 07 m 0 s, 271d 45′ 48″; Dec: −16d 14′ 44″; Diameter: 0d 05′)

Trail of light without stars, 5–6 min in coverage and spindle-shaped, and is nearly like the belt of Andromeda but with a very faint light; there are two telescopic stars close by and placed parallel to the equator. In a clear sky this nebula can be seen very well with an ordinary telescope of 3 ft & a half. Reviewed 22 March 1781.

How to View

Whilst large apertures are certainly an advantage for M17, instruments from about 100 mm aperture upwards show some detail. Its small size also means that magnifications of up to 150x will show more detail in medium sized instruments. From light-polluted skies, a light pollution reduction (LPR) filter allows more detail to be shown.

Photographic Details

Figures 3.48 and 3.50

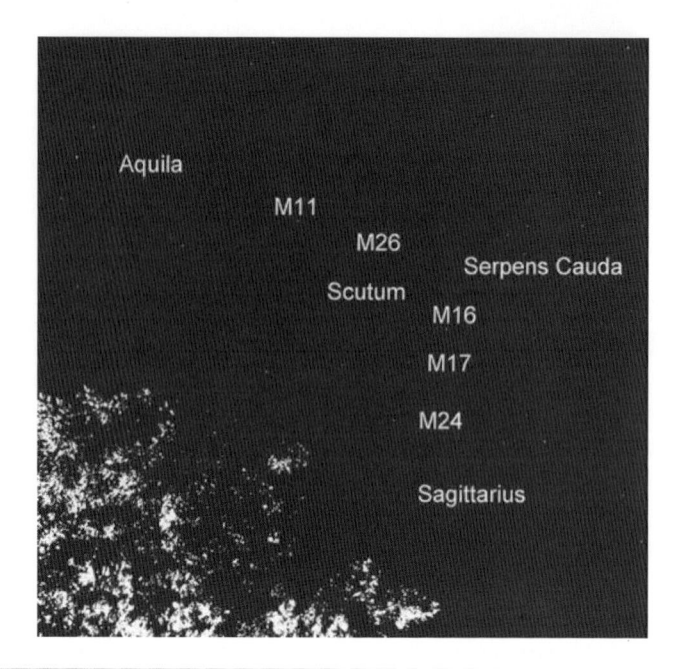

Fig. 3.49 How to find M17

Fig. 3.50 M17 modified

M18

M18 is an open star cluster in the southern part of the sky towards the galactic center. It is about a fifth of a degree across but only has 20 members readily visible to small amateur instruments. It is in the same part of the sky as M16 and M17 but, being an open star cluster, does not suffer the same degree of extinction as its more famous neighbours. It was discovered by Charles Messier in 1764. As it is a sparse cluster, it is hard to see how it could have been confused with a comet. It is 9 arcminutes across and about 4,900 light years away.

Where to Find

If M18 was in a sparse part of sky, it would be tough to find. Fortunately, it isn't and it is in the same binocular field of view as the much easier M24 (Fig. 3.51).

Fig. 3.51 M18 (NASA/courtesy of nasaimages.org)

Object Type	Open star cluster
Declination	−17° 08 min
Right Ascension	18 h 19.9 min
Magnitude	7.5

What It Looks like

M18 appears as a small, sparse open star cluster to 70 mm binoculars and it would be very tempting to cross it off quickly while enjoying its more interesting neighbors. The effort in moving to a medium aperture instrument with light pollution reduction filters is well worth it. Being bang in the Milky Way means that there are a host of fainter stars in the field of view which, although not true cluster members, nevertheless make it an interesting view.

Charles Messier's Original Notes

3 June 1764
(RA: 18 h 06 m 16 s, 271d 34′ 03″; Dec: −17d 13′ 14″; Diameter: 0d 05′)

Cluster of small stars, a little below the nebula No. 17 above, surrounded by slight nebulosity, this cluster is less apparent than the last, No. 16: with an ordinary telescope of 3 ft and a half, this cluster appears as a nebula, but with a good telescope one sees only stars.

How to View

Strangely enough, M18 is one of the few objects where large apertures do not enhance the view. However, larger apertures do bring out more of the background Milky Way stars but, fortunately, the cluster does not get lost amongst them.

Photographic Details

This is already a good representation of how M18 looks through a 127 mm Maksutov, so needs no modification (Fig. 3.51).

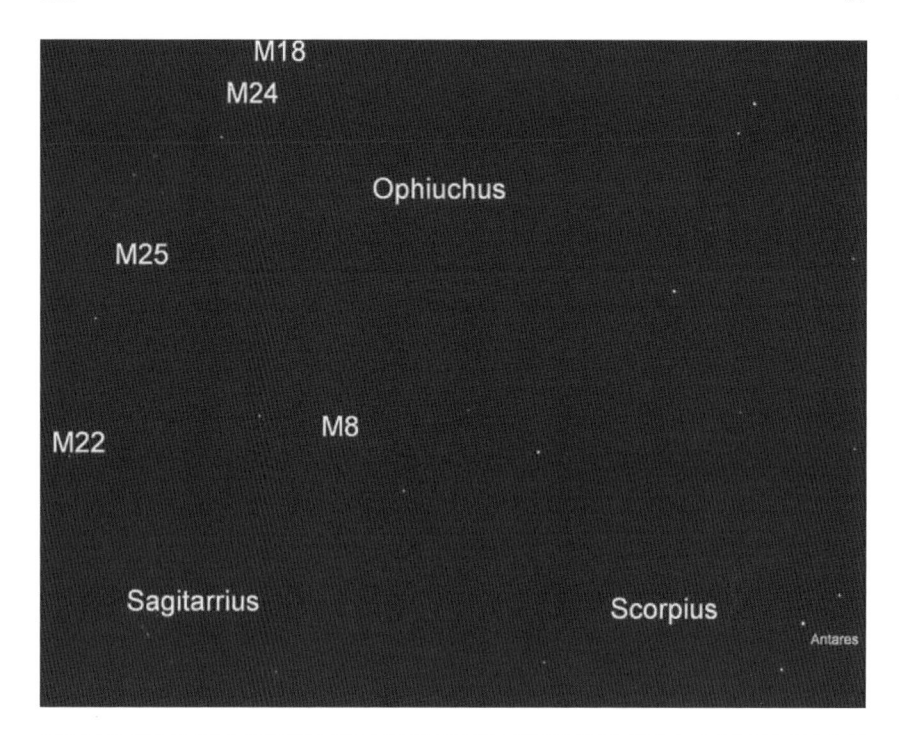

Fig. 3.52 How to find M18

M19

M19 is yet another globular star cluster in the general direction of the galactic center. It is unusual in that it is very elliptical being about 17×9 arcminutes, although only the central 6 arcminutes are readily visible to amateur instruments. Scientists speculate that its unusual shape is due to its proximity to the galactic center of "only" about 5,200 light years, while its about 28,000 light years from us. Its southerly declination makes it a very difficult target from UK and Canadian skies and, even at more convenient latitudes, medium aperture amateur instruments are unable to see anything more than an elliptical patch, as its brightest stars are magnitude 14. It was discovered by Charles Messier in 1764.

Where to Find

M19 is in a rather spare part of sky, although against the background of the Milky Way. It is about 5° east of Antares, although the sharp-eyed may sweep it up in binoculars (Fig. 3.53).

Fig. 3.53 M19 (NASA/courtesy of nasaimages.org)

Object Type	Globular star cluster
Declination	−26° 16 min
Right Ascension	17 h 02.6 min
Magnitude	7.1

What It Looks like

M19 is just about visible in 70 mm binoculars on a clear night. Through a 127 mm Maksutov, it is easier to make out but there is little to suggest its true nature. It is interesting to compare it with nearby M62, which despite being fainter, has a noticeable core.

Charles Messier's Original Notes

5 June 1764
(RA: 16 h 48 m 07 s, 252d 01' 45"; Dec: −25d 54' 46"; Diameter: 0d 03')

Nebula without stars, on the parallel of Antares between Scorpius and the right foot of Ophiuchus: this nebula is round; one can see it very well with an ordinary telescope of 3 ft & a half; The closest known star to this nebula is 28 Ophiuchii, of 6th magnitude, according to Flamsteed. Reviewed 22 March 1781.

How to View

Although visible in small instruments, a large aperture is definitely an advantage. As it has a low southerly declination, 300 mm aperture is needed to resolve the brighter component stars. A light pollution reduction filter offers little help with small aperture telescopes and binoculars.

Photographic Details

Figures 3.54 and 3.55

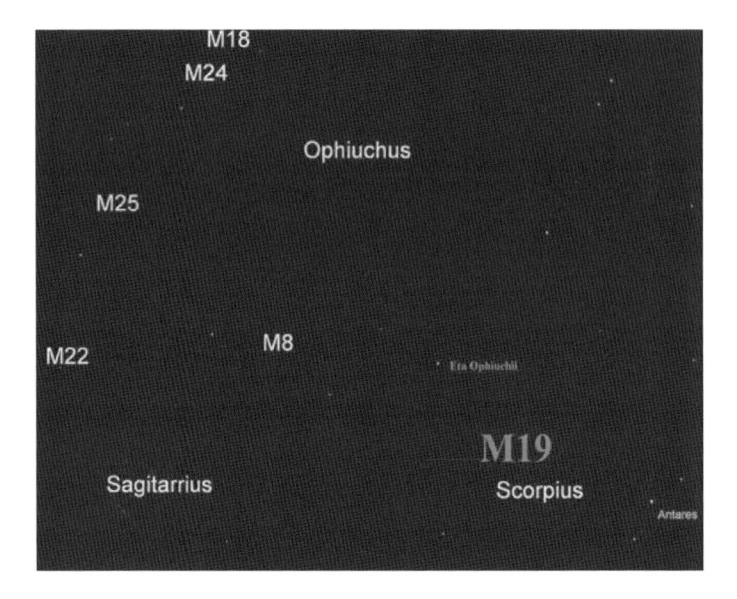

Fig. 3.54 How to find M19

Fig. 3.55 M19 (Modified NASA/courtesy of nasaimages.org)

M20

M20 is known as the Triffid Nebula. It was discovered by Charles Messier in 1764. At 28 arcminutes across, it is almost as large as the Moon as seen from Earth. From the UK and Canada, it is a very difficult object and Messier's original description only hinted at the existence of nebulousity surrounding a cluster of stars. It is likely that extinction would have been a problem for him, even at the more southern latitudes of France. Larger aperture instruments show its unusual shape is due to the presence of an emission nebula, reflection nebula and dark nebula. It is about 5,200 light years away, according to some estimates and nearly 9,000 light years away, according to others.

Fig. 3.56 M20 (Courtesy of Joe Billington, Manchester Astronomical Society)

Where to Find

M20 would be very hard to find were it not for its close proximity to M8. With an instrument with a wide field of view, it is possible to get parts of both objects into the same field of view, although the centers of each are about a degree and a half apart (Fig. 3.57).

Object Type	Nebula
Declination	−23° 02 min
Right Ascension	18 h 02.6 min
Magnitude	7.6

What It Looks like

M20 is too faint to be detected in 70 mm binoculars from northern temperate latitudes. However, as it is close to M8, it is easy to find in a 127 mm Maksutov. In many ways it is similar to M8, showing both nebulosity and quite a few stars. However, the nebulosity will be missed on all but the clearest of nights. Were it not so close to M8, it would probably get more attention. As it is a much photographed object, the visual impression can be disappointing as you miss the color contrast and dark lanes. It is worth persevering with and is an object to be visited and revisited, rather than be rushed through on a "let's bag as many Messier objects as one can" session.

Charles Messier's Original Notes

5 June 1764
(RA: 17 h 48 m 16 s, 267d 04' 05"; Dec: −22d 59' 10"; Diameter: −)

Star cluster, a little above the Ecliptic, between the bow of Sagittarius & the right foot of Ophiuchus. Reviewed 22 March 1781.

How to View

M20 is rather faint and its brightness is spread out over a large area. This makes it tough for medium aperture instruments, so it is best to use low magnification and a light pollution reduction filter. Too much magnification and the light becomes spread as to make it invisible. At the risk of repeating a cliché, aperture really helps.

Photographic Details

Figures 3.56 and 3.58

M20   93

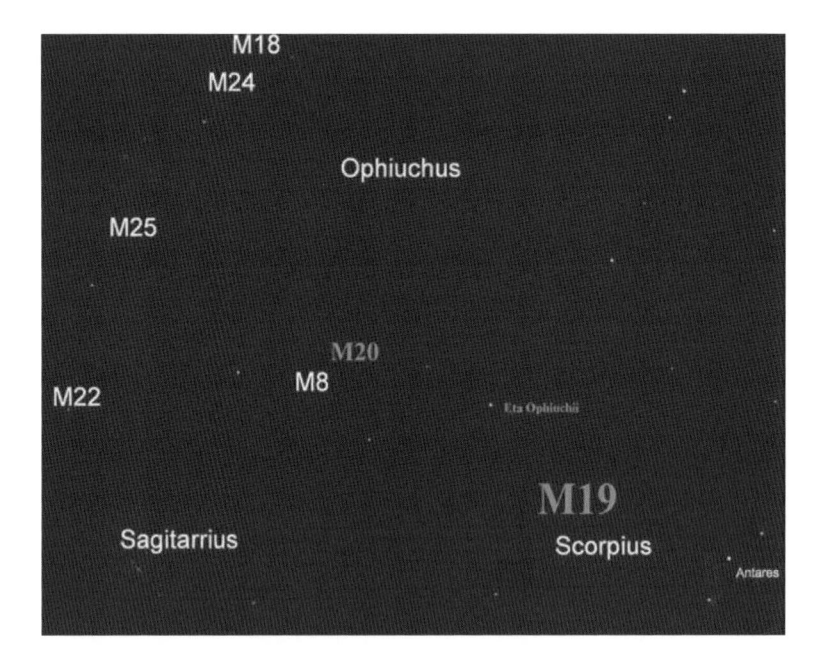

Fig. 3.57 How to find M20

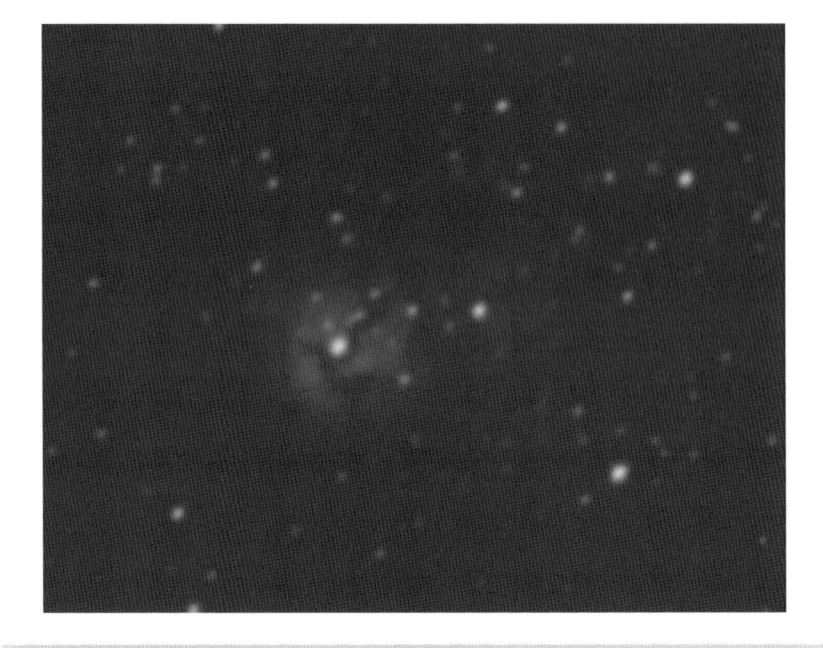

Fig. 3.58 M20 modified

M21

M21 is an open star cluster in Sagittarius. It is quite close to the Triffid Nebula (M20) and it is possible that the two could be physically associated. Indeed, the experts cannot agree on its distance from us and it may be nearer or further than M20. It was also discovered by Charles Messier. Its southerly declination means that it isn't seen very often from the UK and Canada, although as it is an open star cluster, it suffers less from extinction than M20 or any of the nearby globular star clusters. It has 50 stars spread over 13 arcminutes, so is quite a sparse cluster, so not a particularly exciting object.

Where to Find

M21 would be a tough object to find, were it not for its proximity to M8 and M20. It is almost exactly 2° north of M8 (Fig. 3.59).

Fig. 3.59 M21 (NASA/courtesy of nasaimages.org)

M21

Object Type	Open star cluster
Declination	−22° 30 min
Right Ascension	18 h 04.6 min
Magnitude	6.5

What It Looks like

M21 should be visible in 70 mm binoculars but it was impossible on an apparently clear night but can be spotted on others. However, in a 127 mm Maksutov, it was clear and seemed to consist of an area of faint stars, with a couple of concentrations which appeared brighter. It was certainly not a memorable object but is worth making repeat visits to, as once you've bagged M8 and M20, as its not too far away.

Charles Messier's Original Notes

5 June 1764
(RA: 17 h 50 m 07 s, 267d 31′ 35″; Dec: −22d 31′ 25″; Diameter: –)

Star cluster, near the preceding; closest known star to these two clusters is 11 Sagittarii, 7th magnitude, according to Flamsteed. The stars of these two clusters are of 8th and 9th magnitude, surrounded by nebulosity.

How to View

It seems that M21 is one of the more difficult objects, despite its integrated magnitude of 6.5. It is spread over a large area, so small to moderate magnifications are best. Whilst larger aperture instruments show more background stars, the sparseness of the cluster suggests that telescopes in the 100–150 mm aperture range are best. A light pollution reduction filter can help on hazy nights but also has a tendency to obscure the fainter members.

Photographic Details

This is a good representation of how M21 looks through a 127 mm Maksutov, so no modification is required (Fig. 3.61).

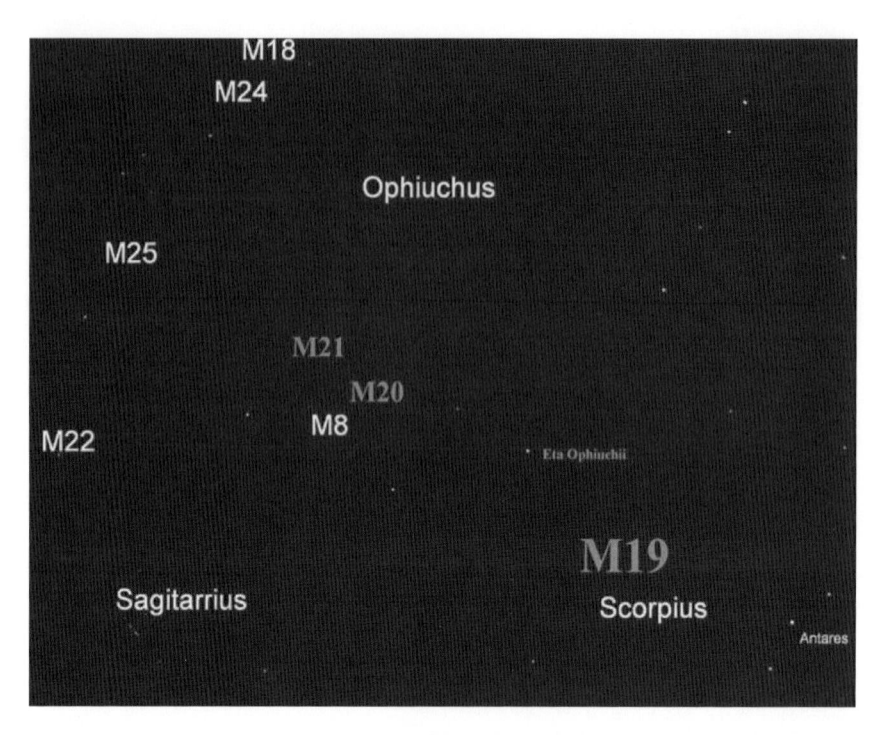

Fig. 3.60 How to find M21

M22

M22 is a large globular star cluster and is one of the nearest ones at about 10,000 light years. It has almost exactly the angular size of the Moon and, were it not for its extreme southerly declination, it would be readily visible to the unaided eye from most parts of the northern hemisphere. It was recorded by Abraham Ihle in 1665 but may have been know earlier than this. Like the better-known M13, it has been studied by researchers and many objects with about 80 Earth masses have been discovered. These are probably too large to be planets, so are probably brown dwarves. As its brightest stars are about magnitude 11, the outer members can be resolved by medium aperture instruments.

Where to Find

Unfortunately, M22 doesn't have any nearby landmarks. The nearest is M8, which is about 7° due west of M22. Fortunately, it is bright enough to be easily visible in binoculars on a clear night (Fig. 3.62).

M22

Fig. 3.61 M22 (Courtesy of Ray Grover, Manchester Astronomical Society)

Object Type	Globular star cluster
Declination	−23° 54 min
Right Ascension	18 h 36.4 min
Magnitude	5.1

What It Looks like

If you have read this book serially, you will have gained the impression that most globular star clusters in the Messier Catalogue are amongst the less exciting objects in it. Indeed, it is true that most of them are not particularly memorable. Yet, it is refreshing to find an exception. M22 is far enough south to suffer from extinction but if you catch it on a very clear night, you can see it very clearly with 70 mm binoculars. The only disappointment is that a 127 mm Maksutov doesn't vastly improve the view and it appears more diffuse. What is nice is that M22 is quite clearly a globular star cluster and the only object type it can be possibly confused with is an elliptical galaxy.

While researching this book, it was definitely a pleasant surprise.

Charles Messier's Original Notes

5 June 1764
(RA: 18 h 21 m 55 s, 275d 28' 39"; Dec: −24d 06' 11"; Diameter: 0d 06')

Nebula, below the Ecliptic, between the head & bow of Sagittarius, near a star of 7th magnitude, 25 Sagittarii, according Flamsteed. This nebula is round, contains no star, & we see it very well with an ordinary telescope of 3 ft & a half; the star λ Sagittarii was used for determination [of position]. Abraham Ihle, German, discovered it in 1665 when observing Saturn. M. Le Gentil observed it in 1747, & made an engraving of it. *Memoirs of the Academy for the year 1759, page 470.* Reviewed 22 March 1781; it is reported in the *English Atlas*.

How to View

M22 appears to be one of the few Messier objects that aren't improved much by aperture. Larger aperture instruments resolve some of the outer stars but at the expense of being able to see the whole object. High magnification does not help but a light pollution reduction filter certainly does.

Photographic Details

Figure 3.61 and 3.63.

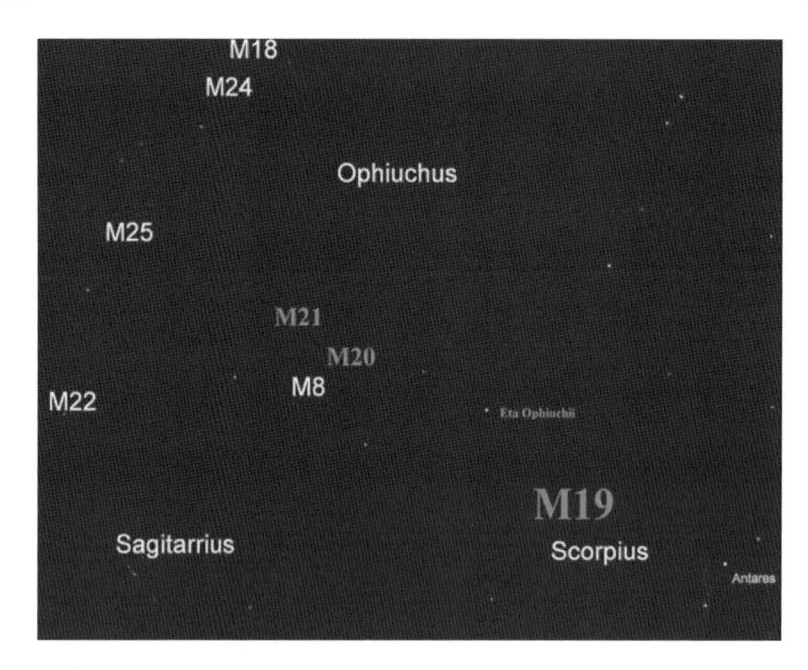

Fig. 3.62 How to find M22

Fig. 3.63 M22 (Courtesy of Ray Grover, Manchester Astronomical Society)

Chapter 4

M23–M45

M23

M23 is an open star cluster in Sagittarius, discovered by Charles Messier in 1764. It has about 150 stars spread across a diameter of about 27 arcminutes and can sometimes be difficult to detect against the background Milky Way. The brightest stars are about 9th magnitude and about 50 stars are 12th magnitude or brighter. Its southerly declination means that it never reaches more than 18° above the horizon from the UK. This is easier than some of the more southern members of the Messier Catalogue but it still needs a clear night to get a good view with modest equipment. It is about 2,150 light years away and there is a foreground star of magnitude 6.9 that is not a member of the cluster.

Where to Find

M23 doesn't have any nearby stars to act as guideposts but is in a generally rich part of the sky. It is about 6° north of M20 and about a degree to the west (Fig. 4.2).

Object Type	Open star cluster
Declination	−19° 01 min
Right Ascension	17 h 56.8 min
Magnitude	6.9

Fig. 4.1 M23 (NASA/courtesy of nasaimages.org)

What It Looks like

Through 70 mm binoculars, M23 looks rather sparse but is still visible under average conditions. Through a 127 mm Maksutov, there seems to be a horseshoe shape which actually appears in the photograph below, along with other stars. Although not one of the more classical open star clusters, it is nevertheless well worth a look.

Charles Messier's Original Notes

20 June 1764
(RA: 17 h 42 m 51 s, 265d 42′ 50″; Dec: −18d 45′ 55″; Diameter: 0d 15′)

M23

Star cluster, between the end of the bow of Sagittarius & the right foot of Ophiuchus, very near the star 65 Ophiuchi according Flamsteed. The stars of the are very close to one another. Its position having been determined via μ Sagittarii.

How to View

Although small aperture instruments are capable of showing the main asterism of M23, a larger aperture instrument with a wide field of view is much better.

Photographic Details

This is a good representation of how M23 looks like through a 127 mm Maksutov, so no modification is required (Fig. 4.1).

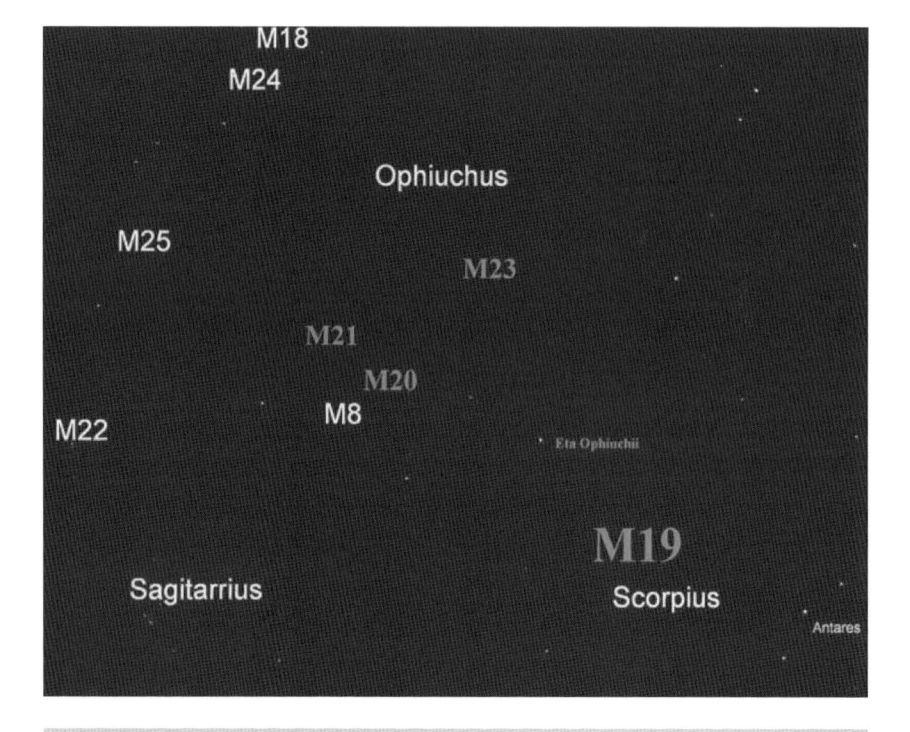

Fig. 4.2 How to find M23

M24

M24 is rather unlike the other members of the Messier Catalogue in that it is really part of the Milky Way. However, when Charles Messier discovered it in 1764, its true nature wasn't known. We now know it to be an arm of our local galaxy 10,000 light years away that appears detached from the rest of it, as it is surrounded by dark nebulae. Its appearance through binoculars is a large misty patch a degree and a half in diameter, which would suggest some sort of star cluster or galaxy.

Where to Find

Like many objects in this region, M24 has no obvious signposts but, honestly, it is prominent enough to be a signpost in itself. It is quite unmistakable in even small instruments (Fig. 4.4).

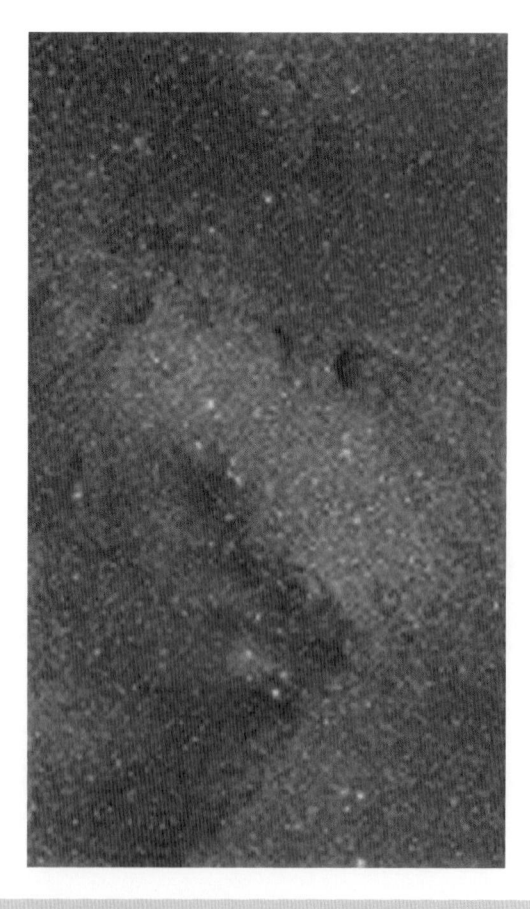

Fig. 4.3 M24 (NASA/courtesy of nasaimages.org)

M24

Object Type	Star cloud
Declination	−18° 29 min
Right Ascension	18 h 16.9 min
Magnitude	4.6

What It Looks like

One phrase sums it up: "eye candy"! Purists would dismiss it as not a "real" deep sky object but what we're really about as amateurs is enjoying it. This is no mere faint fuzzy to cross off the list. Its size and official magnitude would suggest a faint, diffuse object but if you've seen Omega Centauri through a small instrument, you'll understand what I mean. It's attractive enough in binoculars but a 127 mm Maksutov armed with suitable accessories will suggest nebulosity which we know are really concentrations of faraway stars. Never mind! Just enjoy the view.

Charles Messier's Original Notes

20 June 1764
(RA: 18 h 01 m 44 s, 270d 26′ 00″; Dec: −18d 26′ 00″; Diameter: 1d 30′)

Cluster on the [same] parallel of the previous [nebula] & near the end of the bow of Sagittarius, in the Milky Way: a big nebula in which there are many stars of different magnitudes: the light that is prevalent in this cluster is divided into several parts; it is the middle of this cluster that has been determined.

How to View

M24 is actually one of the easier Messier objects and it can be enjoyed through modest instruments. Aperture works but only to a certain extent. You really need about 2° field of view to enjoy it properly, so this limits telescopes to about 800 mm focal length, although focal reducers can be used to get around this restriction. A light pollution reduction filter helps, too.

Photographic Details

Figures 4.3 and 4.5

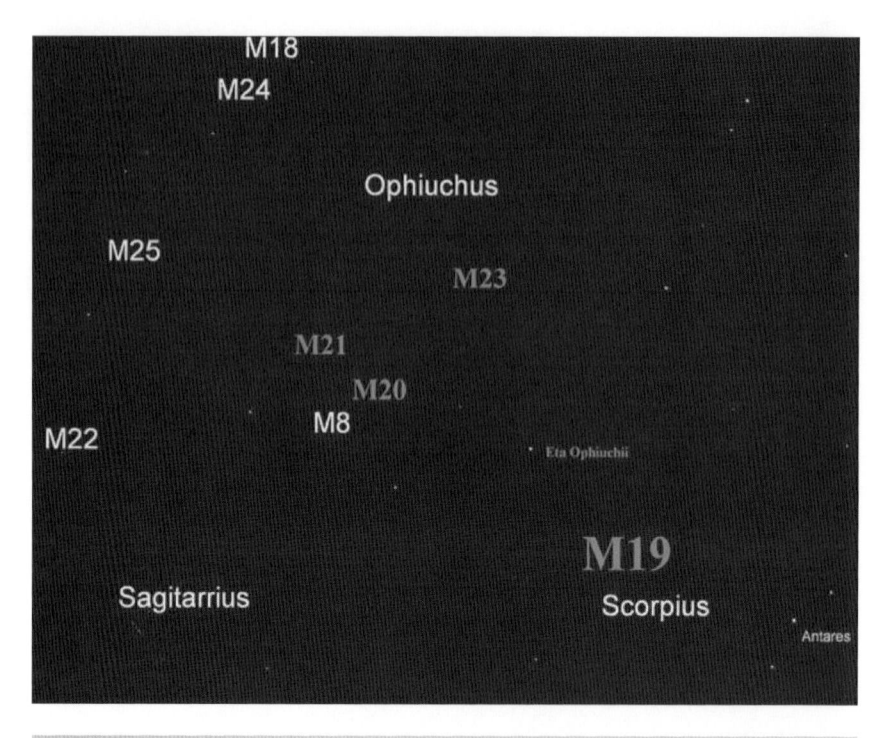

Fig. 4.4 How to find M24

M25

M25 is an open star cluster in Sagittarius. As it is in a particularly rich part of the Milky Way, experts do not agree which stars are cluster members and which are merely "field" stars. Most distance estimates of about 2,000 light years and size estimates of 32 arcminutes seem to agree.

It is not at its best when seen from mid-northern latitudes but can be picked up easily enough in binocular sweeps on a moderately clear night. It was discovered by Philippe Loys de Chéseaux in 1745 but being relatively bright, may have been known about earlier.

Where to Find

The bad news is that there's no easy signposts to M25 but, fortunately, it is bright and large and, when due south is surprisingly high in the sky, getting about 18° clear of the horizon in southern England and Canada (Fig. 4.7).

M25

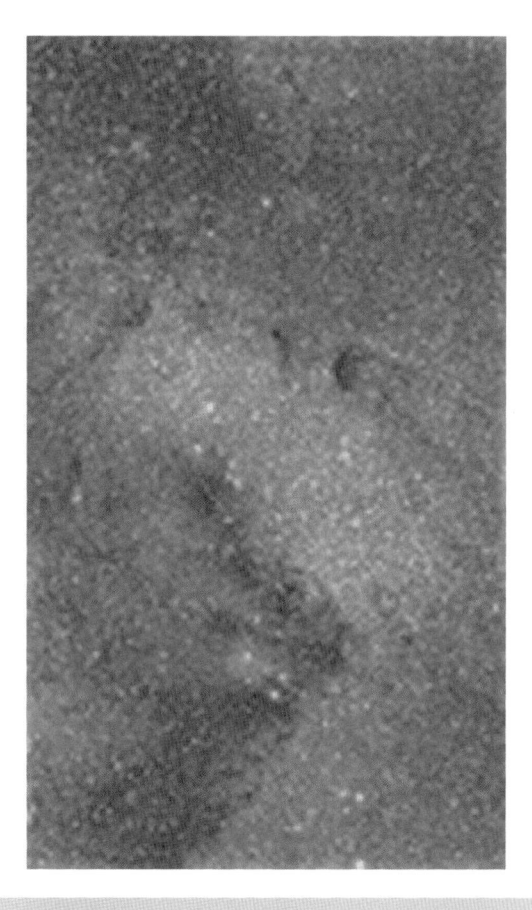

Fig. 4.5 M24 (Modified NASA/courtesy of nasaimages.org)

Object Type	Open star cluster
Declination	−19° 15 min
Right Ascension	18 h 31.6 min
Magnitude	4.6

What It Looks like

In 70 mm binoculars, M25 looks like an open star cluster (hardly surprising) and a large and rich one but not matching the better known Pleiades (M45) for impact. It is well worth observing it in a larger instrument (a 127 mm Maksutov for example). There are an uncountable number of stars, although there is some doubt as to which are true members. There also appears to be a hint of nebulosity but I suspect it is an illusion caused by dense clumps of background stars and fainter cluster members.

Fig. 4.6 M25 (NASA/courtesy of nasaimages.org)

Charles Messier's Original Notes

20 June 1764
(RA: 18 h 17 m 40 s, 274d 25′ 00″; Dec: −19d 05′ 00″; Diameter: 0d 10′)

Cluster of small stars in the vicinity of the previous two clusters, between the head and the tip of the bow of Sagittarius: the closest known star to this cluster is 21 Sagittarii, 6th magnitude according Flamsteed. The stars of this cluster are difficult to see with a regular telescope of 3 ft and one sees no nebulosity. Its position has been determined from the star μ Sagittarii.

How to View

M25 is actually one of the easier Messier objects and it can be enjoyed through modest instruments. You really need about 45 arcminutes field of view to enjoy it properly,

M26 109

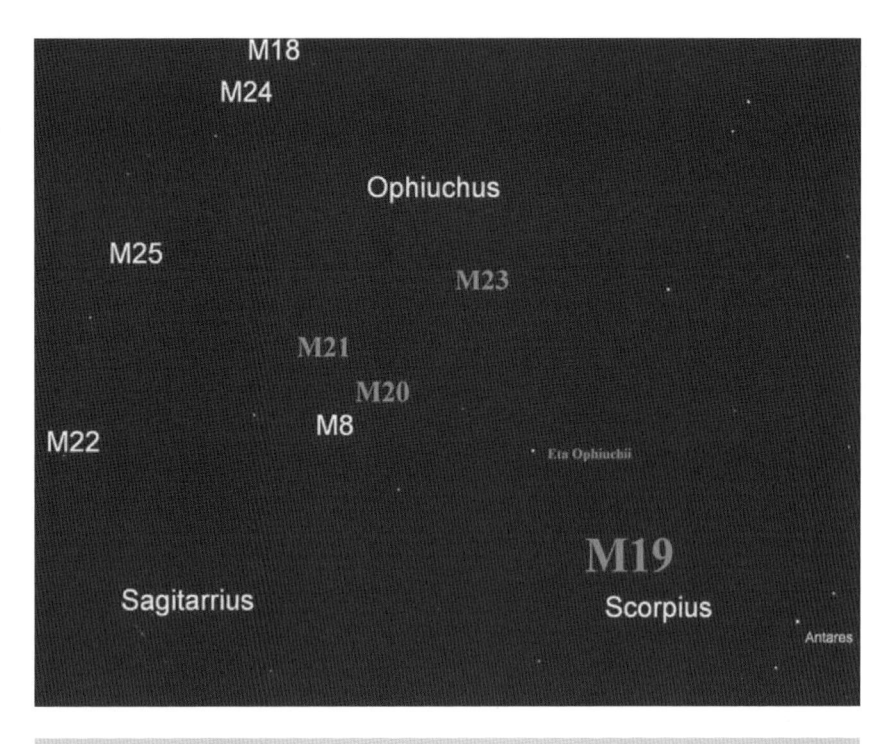

Fig. 4.7 How to find M25

so this limits telescopes to about 2,400 mm focal length, which is not a problem for most large amateur telescopes. A light pollution reduction filter helps, too.

Photographic Details

This is a good representation of how M25 appears in a 127 mm Maksutov, so needs no modification (Fig. 4.6).

M26

M26 is an open star cluster in Scutum. It was discovered by Charles Messier in 1764 and is rather faint, with its brighter stars a bit brighter than 12th magnitude. This means that it will be missed by casual observers with medium aperture instruments. It is 15 arcminutes across and has 25 stars visible in 150–200 mm instruments and about 100 altogether. It is about 5,000 light years away.

Fig. 4.8 M26 (NASA/courtesy of nasaimages.org)

Where to Find

M26 is best found in binoculars by sweeping from Altair to M11 and beyond. However, you need to sweep slowly, as it is quite faint (Fig. 4.9).

Object Type	Open star cluster
Declination	−9° 24 min
Right Ascension	18 h 45.2 min
Magnitude	8.0

What It Looks like

M26 is overshadowed by its more stellar neighbors (if you forgive the pun!). Indeed, the only entry in my observing log in binoculars is "faint". Indeed, it is small wonder that Charles Messier was not aware of its true nature. Some of the brighter stars become visible in a 127 mm Maksutov but it has a sparse and not particularly pleasing appearance. It really needs a much bigger aperture.

Charles Messier's Original Notes

20 June 1764
(RA: 18 h 32 m 22 s, 278d 05' 25"; Dec: −09d 38' 14"; Diameter: 0d 02')

Star cluster near [the stars] η & o in Antinous, between which there is one which has more brightness: with a telescope of 3 ft one cannot distinguish them, one must employ a good instrument. The cluster contains no nebulosity.

How to View

Although M26 can be seen in smaller instruments, it really needs a large aperture of at least 200 mm to fully appreciate it. With it being sparse, low magnification is the best way to see it.

Photographic Details

This is a good representation of how M26 appears in a 127 mm Maksutov, so needs no modification (Fig. 4.8).

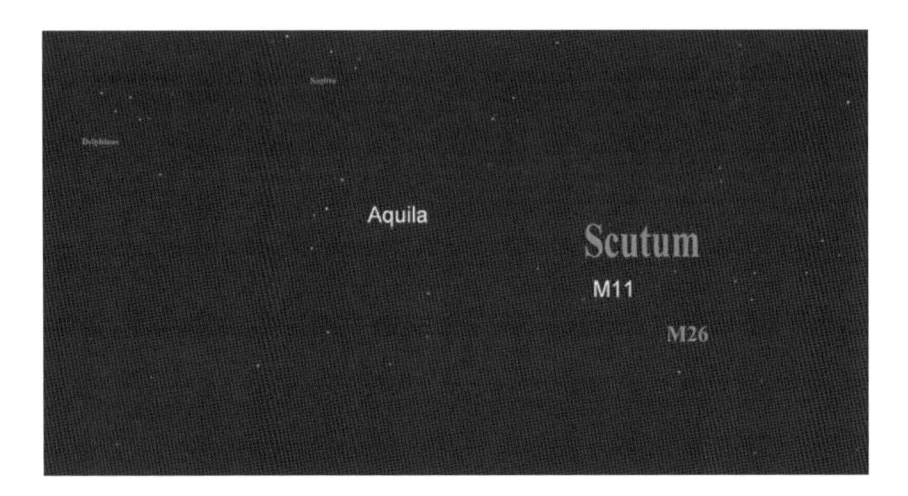

Fig. 4.9 How to find M26

M27

M27 is also known as the Dumbbell Nebula. It is the brightest planetary nebula north of the celestial equator and the first ever discovered. Charles Messier discovered its existence in 1764, although its true nature wasn't discovered until later. It has an angular extent of 8×5.7 arcminutes and there are several distance estimates between 490 light years and 3,500 light years. This means that its size and apparent magnitude cannot be determined. What is known is that the progenitor star was Sun-like (between 0.2 and 3 solar masses) and it shed its outer layers while changing from a red giant into a white dwarf.

Its high northerly declination means it is well-placed in the northern hemisphere during summer, reaching a healthy 52° elevation from northern Scotland and northern Canada. It was visible the first night my Maksutov-Cassegrain arrived in 2002 and it was amazing how close it appeared visually to the many photographs on the net. People don't normally see it as often as the Ring (M57), as it is harder to find, even though it is brighter.

Fig. 4.10 M27 (photographed by Anthony Glover)

Where to Find

Although M27 is the brightest object of its type, its position in the faint constellation of Vulpecula makes it harder to find than the fainter but better known Ring Nebula (M57) in Lyra. It is about 3° north of Gamma Sagittae, the easternmost star in Sagitta. Refer to the description of M71 to find Sagitta (Fig. 4.11).

Object Type	Planetary nebula
Declination	22° 43 min
Right Ascension	19 h 59.6 min
Magnitude	7.4

What It Looks like

Although M27 is not one of the easiest Messier objects to find without electronic aid (which was not available to Charles Messier), it is well worth the view when you see it. It is larger and brighter than the Ring (M57) and its structure is obvious in a 127 mm Maksutov. In more modest instruments (such as 70 mm binoculars) it appears as a fuzzy patch of indeterminate nature but larger instruments bring out its structure as two lobes connected by a central point. If you can imagine Fig. 4.11

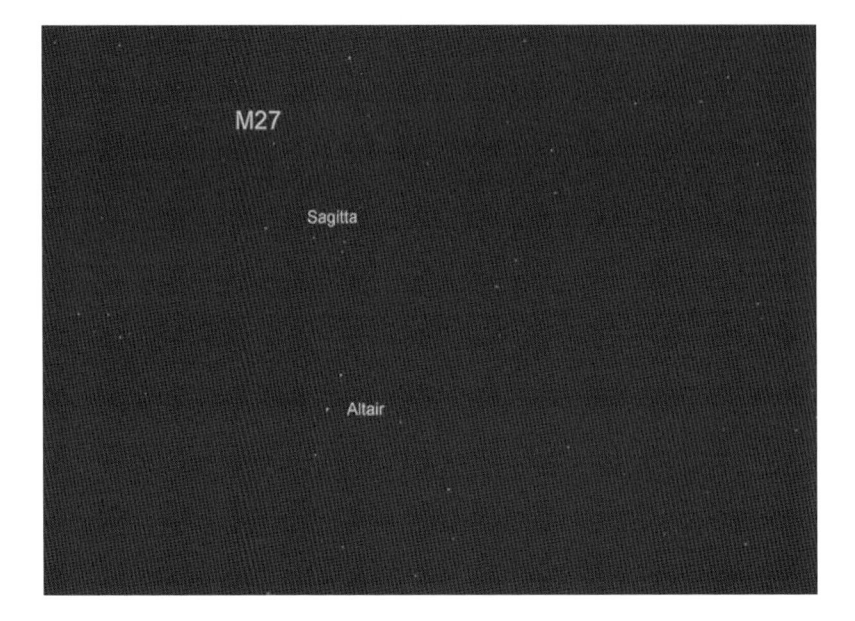

Fig. 4.11 How to find M27

without the outer envelope and in monochrome, it's a very good approximation to how it looks through a 127 mm Maksutov on a good night. It is definitely one of the more interesting and visually pleasing Messier objects.

Charles Messier's Original Notes

12 July 1764
(RA: 19 h 49 m 27 s, 297d 21' 41"; Dec: +22d 04' 00"; Diameter: 0d 04')

Nebula without star, discovered in the Fox [Vulpecula] between the two front paws, and very near the star 14 of this constellation, of 5th magnitude according to Flamsteed; we see it with a regular telescope of 3 ft and a half: it appears as an oval shape, and contains no star. M. Messier has reported its position on the chart of the Comet of 1779, which will be engraved for the volume of the Academy in that same year. Reviewed 31 January 1781.

How to View

M27's modest size and brightness make it a good candidate for using high magnification, unlike many Messier objects. Large instruments of long focal length are not a disadvantage until the field of view becomes much less than 12 arcminutes. It can be seen in large binoculars and small telescopes but really needs an aperture of at least 100 mm to do it justice and reveal its true nature.

Photographic Details

Figure 4.10 was taken using a 110 mm F/7 APO and SXV H9 CCD. Ten 5 min exposures were taken. However, the view through a 127 mm Maksutov shows only slightly less detail and fewer background stars (Fig. 4.12).

Fig. 4.12 M27 modified

M28

M28 is another globular star cluster in Sagittarius, discovered by Charles Messier in 1764. It is about 18,000 light years away, so quite close to the galactic center. As its individual stars are rather faint, it appears like an elliptical galaxy about 11 arc-minutes across. As it is faint and has a low southerly declination, it is never seen at its best from the UK and Canada, suffering a lot from extinction.

Where to Find

M28 is quite near to Lamda Sagittarii (not visible in the above photo) but one way to find it is to draw an imaginary line joining M8 and M22. Move 2° from M22 west towards M8, then move a fraction south (Fig. 4.14).

Fig. 4.13 M28 (NASA/courtesy of nasaimages.org)

Object Type	Globular star cluster
Declination	−24° 52 min
Right Ascension	18 h 24.5 min
Magnitude	7.2

What It Looks like

Although theoretically visible in 70 mm binoculars, M28 was beyond their reach on a clear night. Indeed, the nearby M22 is far more interesting. It was visible to a 127 mm Maksutov but only as a "faint fuzzy" of indeterminate type.

Charles Messier's Original Notes

27 July 1764
(RA: 18 h 09 m 58 s, 272d 29′ 30″; Dec: −24d 57′ 11″; Diameter: 0d 02′)

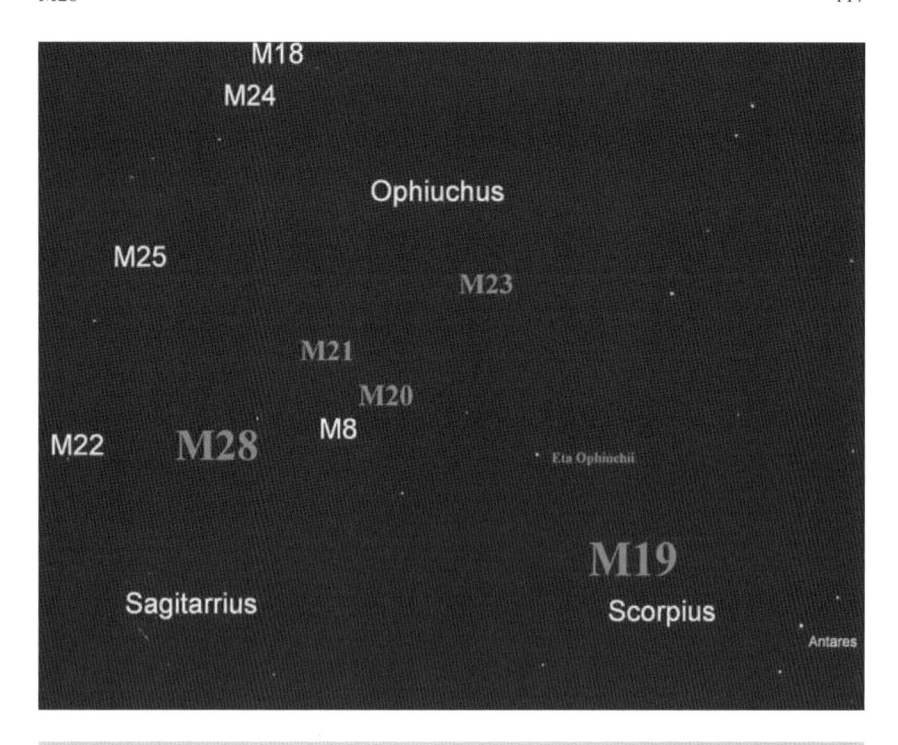

Fig. 4.14 How to find M28

Nebula discovered in the upper part of the bow of Sagittarius one degree from the vicinity of the star λ & little distant from the beautiful nebula which is between the head and the bow. It contains no star, it is round, it can hardly be seen with an ordinary telescope of three and a half feet. Its position was determined from λ Sagittarii. Reviewed 20th March 1781.

How to View

Although M28 is visible in a 127 mm Maksutov, it needs a larger aperture telescope to discern that it is a globular star cluster.

Photographic Details

Figures 4.13 and 4.15.

Fig. 4.15 M28 (Modified NASA/courtesy of nasaimages.org)

M29

M29 is an open star cluster in Cygnus. It was discovered by Charles Messier in 1764 and is 7 arcminutes across. Estimates of its distance vary from 4,000 to 7,000 light years, as there is a lot of interstellar material that has a dimming effect on it, similar to the extinction caused by the Earth's atmosphere to objects near the horizon. Its brightest stars are 8th magnitude, putting the cluster into binocular range, and it is possible that their absolute brightness may exceed stellar beacons such as Deneb and Rigel. Its northerly declination makes it circumpolar from most of the UK and parts of Canada. During summer, it is near the zenith from the northern hemisphere and one of the causes of strained necks from binocular users. On very clear nights, it can be easily lost against the rich Milky Way background and "less than perfect" nights are best to view it.

Fig. 4.16 M29 by Anthony Glover

Where to Find

On a really clear night, M29 can actually be very hard to find against the Milky Way background but it is about a degree and a half south of Gamma Cygni, so can be seen in the same binocular or finderscope field of view (Fig. 4.17).

Object Type	Open star cluster
Declination	38° 32 min
Right Ascension	20 h 23.9 min
Magnitude	7.1

What It Looks like

Were it placed outside of the plane of the Milky Way, M29 would look spectacular. On really clear nights, binocular scans miss it against the rich background. In a good finderscope, there are loads of star fields and finding which one is the real

M29 needs a telescope, such as a Maksutov. There is a main asterism of stars, many of which form a rectangle but which stars are real members and which are background is somewhat of a mystery.

Charles Messier's Original Notes

29 July 1764
(RA: 20 h 15 m 38 s, 303d 54′ 29″; Dec: +37d 11′ 57″; Diameter: –)

Clusters of seven or eight very small stars, which are below γ Cygni, viewed through an ordinary telescope of 3 ft and a half in the form of a nebula. Its position determined from γ Cygni. This cluster is reported on the chart of the Comet of 1779.

How to View

On really clear nights, aperture can work against you, as it brings more Milky Way background stars into view. It is best to wait for a slightly hazy night or use a small aperture. An alternative approach, with larger apertures, is to use a higher magnification.

Photographic Details

This is a good representation of how M29 appears in a 127 mm Maksutov, so needs no modification.

Figure 4.16 was taken using a 80 mm F/6 APO and SXV H9 CCD. Twenty 1 min exposures were taken. Detail: Starlight Xpress SXV H9 and WO 80 mm F6 refractor. Twenty 1 min subs IR blocking filter only.

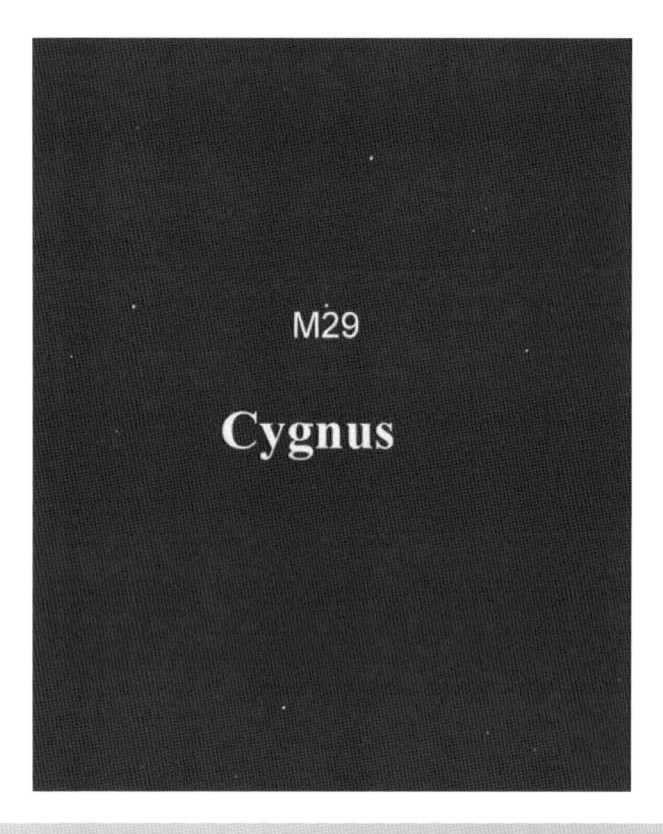

Fig. 4.17 How to find M29

M30

M30 is a globular star cluster in Capricorn, discovered by Charles Messier in 1764. It is 12 arcminutes across but very concentrated towards its core. It is about 26,000 light years away. Its brightest stars are 12th magnitude, bringing them within range of many amateur instruments, although its low southerly declination makes it difficult from the UK and Canada and it isn't at its best even from southern parts of the USA.

Where to Find

At the time of writing, Jupiter was suitably placed as a signpost to find Delta and Gamma Aquarii, with M30 being about 5° south of Gamma (Fig. 4.19).

Fig. 4.18 M30 (NASA/courtesy of nasaimages.org)

Object Type	Globular star cluster
Declination	−23° 11 min
Right Ascension	21 h 40.4 min
Magnitude	7.2

What It Looks like

Only the central region of M30 was visible in 70 mm binoculars on a good night, so it was hard to distinguish it from a star. A 127 mm Maksutov (armed with a light pollution reduction filter) showed it as a globular star cluster but only about half of its published size.

Charles Messier's Original Notes

3 August 1764
(RA: 21 h 27 m 05 s, 321d 46′ 18″; Dec: −24d 19′ 04″; Diameter: 0d 02′)

M30

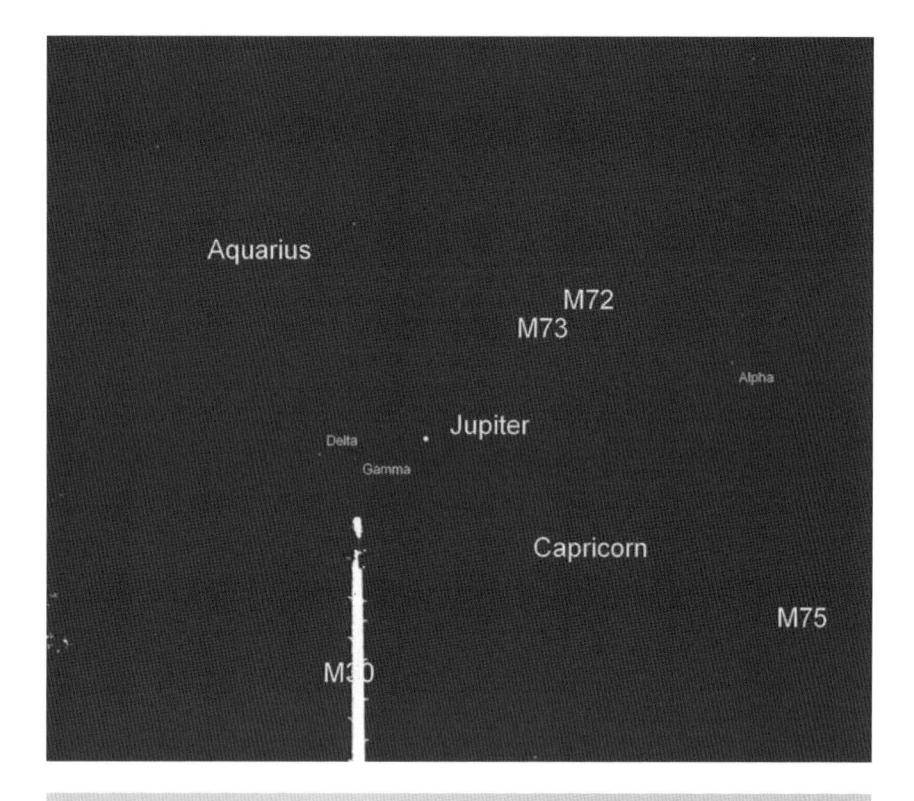

Fig. 4.19 How to find M30

Nebula discovered below the tail of Capricorn, very near the star 41 of this constellation, of 6th magnitude, according to Flamsteed. It is hard to see with an ordinary telescope of 3 ft and a half. It is round & contains no star; its position determined from ζ Capricorni, M. Messier has reported it on the chart of the Comet of 1759. *Memoirs of the Academy for the year 1760, plate, II.*

How to View

M30 is a difficult object from the northern hemisphere and, although its nature can just about be made out in a small instrument, it really needs a larger aperture.

Photographic Details

Figures 4.18 and 4.20

Fig. 4.20 M30 (Modified NASA/courtesy of nasaimages.org)

M31

M31 is sometimes incorrectly referred to as "Andromeda" in popular culture, whilst it is known to astronomers as "The Andromeda Galaxy" or "Andromeda Great Spiral". It has captured the imagination of the scientific community and the general public, in my opinion rightly so. It was known to Al-Sufi in 905 but may well have been known about before. It is a challenge to the unaided eye in modern, light-polluted skies but would have been readily visible in prehistoric times. However, its true nature wasn't even hinted at until William Herschel managed to resolve some of its stars. Vesto Slipher in 1912 and Edwin Hubble in 1923 established that it lay well outside the bounds of our own Milky Way galaxy. Hubble estimated its distance to be 900,000 light years, although most modern estimates place it between 2.2 and 2.7 million light years. This discrepancy is due to a lot of uncertainties about the cosmic scale. The primary method of calculating distance to objects in our Local Group is by examining light variations in Cepheid variables. The longer the period of such a variable, the greater its apparent magnitude (or real brightness). So if you know the period and how bright it appears to be, then you can work out the distance.

Or ... can you? Edwin Hubble was unaware that there were two types of Cepheid variable: Type I and Type II. Indeed, it wasn't until 1953 that we realised that it was much further away than we thought. There is little doubt that M31 is the physically largest galaxy in our Local Group but it appears to be less massive than our own Milky Way, according to some estimates. In about five billion years we will merge with M31 to form a giant elliptical galaxy.

M31 isn't the brightest galaxy in the night sky, as the Clouds of Magellan are much nearer but it is the brightest galaxy visible from north of the equator and certainly the brightest spiral galaxy, outshining is neighbour, M33. M31 has two companion galaxies in the Messier Catalogue, M32 and M110, which can be seen in medium aperture instruments. Certainly Charles Messier knew M32, although M110 was added later.

In larger aperture instruments, even amateur ones, globular star clusters have been seen and photographed in the vicinity. Although small instruments only show the galactic nucleus, even large binoculars are capable of showing is structure. As its angular size is 3×1°, a short tube refractor with a 4° field of view is normally my instrument of choice.

Where to Find

M31 is in quite a sparse part of sky, away from the Milky Way background, so has no bright stars within 2°. Fortunately, it is quite bright, despite being spread out over a large area and has two convenient pointers in Beta and Mu Andromedae. Draw an imaginary line from Beta to Mu and carry on for the same distance. This takes you to M31 (Fig. 4.21).

Object Type	Spiral galaxy
Declination	41° 16 min
Right Ascension	00 h 42.7 min
Magnitude	3.4

What It Looks like

A first view of M31 through modest instruments may be disappointing. It is one of the most photographed objects and these photographs show its spiral structure and even some of its brighter globular star clusters. To the unaided eye, it appears as a misty patch on an exceptional night. With small binoculars, you can tell that there is a brighter, central nucleus and fainter surrounding area. With larger binoculars, the spiral structure becomes evident and when well placed, on nights of exceptional clarity, even seen some dust lanes are apparent, although this is an exception, rather than a regular occurrence. Larger instruments make it easier to detect the satellite

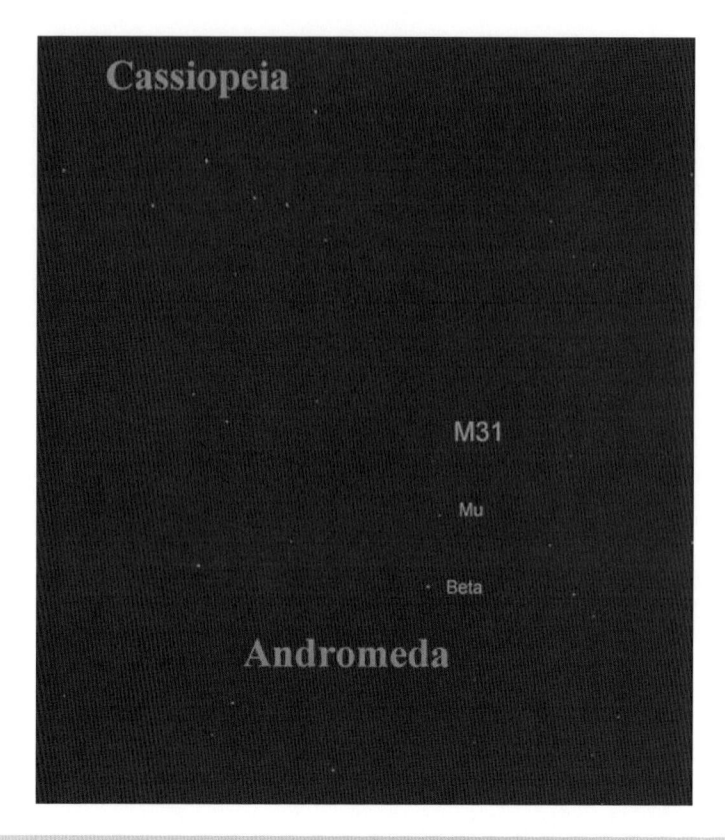

Fig. 4.21 How to find M31

galaxies M32 and M110 but the effect of seeing the whole extent of the galaxy becomes lost, as it is impossible to obtain the necessary field of view. However, dust lanes and variations in the densities of the spiral arms are easier to see.

Charles Messier's Original Notes

3 August 1764
(RA: 00 h 29 m 46 s, 07d 26′ 32″; Dec: +39d 09′ 32″; Diameter: 0d 40′)

The beautiful nebula of the belt of Andromeda, shaped like a spindle; M. Messier has examined it with various instruments, and there is no recognized star: it resembles two cones or pyramids of light opposed at the bases, whose axis is in the north-west to south-east, and the two points of light or the two summits are nearly distant from each other by 40 min of arc; the common base of both pyramids is 15 min [of arc]. This nebula was discovered in 1612 by Simon Marius, and then observed by

M31   127

Fig. 4.22 M31 by Anthony Glover

different astronomers. M. Le Gentil has given a drawing in the *Memoirs of the Academy for the year 1759, page 453*. It is reported in the *English Atlas*.

How to View

On an average night, even small binoculars or a small finderscope will show M31 and it should always be included it on a quick binocular tour of the night sky. The key to getting a good view of it is to use an instrument with a wide field of view, preferably upwards of 4°. It is impressive enough through binoculars in the 50 mm to 70 mm aperture range but apochromatic refractors with a short focal length are the best instruments for seeing and snapping it in its entirety.

Photographic Details

M31 was taken with an 80 mm/F6 APO refractor with a focal reducer/field flattener at 0.8 focal ratio. Ten 5 min exposures were taken using an ST4000×CM color camera (Fig. 4.22).

Fig. 4.23 M31 modified

M31 – Williams Optics 80 mm F6 Apo, SBIG ST4000×cm (NOT the ST2000×cm), TRF2008 0.8 Focal Reducer/Field Flattener, 10×5 min subs (Fig. 4.23).

M32

M32 is an elliptical galaxy, which is a satellite galaxy of M31. It was discovered by Guilliame Le Gentil in 1749. From Earth, it appears to be slightly elliptical, as opposed to circular and 22 arcminutes south of M31's core. It has the same "official" distance as M31 but is believed to be in its foreground of the spiral arms. Were it on its own, it would be quite difficult for small instruments, being only 8×6 arcminutes. Indeed, at first sight it appears like a star but you can soon see the outlying halo and that it isn't quite a point source of light.

It is quite an interesting object scientifically, as it has a dense core surrounding a massive object, almost certainly a black hole. It has been resolved into stars, confirming that it is a distant object and not one of the Milky Way's globular star clusters.

Where to Find

M32 is so close to M31 that it is actually in front of its outer spiral arms, so look for a small galaxy with a bright nucleus in front of its outer regions.

M32

Object Type	Elliptical galaxy
Declination	40° 52 min
Right Ascension	00 h 42.7 min
Magnitude	8.1

What It Looks like

At first glance, M32 just looks like a background star but if you look closely you can see its more diffuse outer regions. Although it is technically an "elliptical" galaxy, it appears nearly circular. Its core is quite compact, compared to most elliptical galaxies, so the overall effect is rather more pleasing than many elliptical galaxies, which appear like a smudge. On a poor night, only the nucleus is visible, so it is easy to mistake for a star.

Charles Messier's Original Notes

3 August 1764
(RA: 00 h 29 m 50 s, 07d 27' 32"; Dec: +38d 45' 34"; Diameter: 0d 02')

Small nebula without stars, below & just minutes [or arc] from the belt of Andromeda; this small nebula is round, its light fainter than the belt. M. Le Gentil discovered it on 29 October 1749. M. Messier saw her for the first time, in 1757, & there is no recognized change.

How to View

If M32 was not close to M31, it would be visible in its own right in large binoculars, although probably only the core. A telescope of at least 100 mm aperture is recommended and also at least a moderately clear night.

Photographic Details

M32 is shown in the widefield shot in Fig. 4.22. Figure 4.24 shows how it looks through the 127 mm Maksutov, with a field of view of about 2° and showing some of M31.

Fig. 4.24 M32 modified

M33

M33 is sometimes known as the Pinwheel or Triangulum Galaxy. With a visual magnitude of 5.7, it suggests that it should be visible to the unaided eye but, as its brightness is spread out over an area of 73×45 arcminutes, it is a far more difficult object than its magnitude would suggest. Although Charles Messier was previously unaware of its existence before compiling his catalogue, it was probably known about before 1654 by Giovanni Battista Hodierna. Claims have been made to have seen it with the unaided eye. Whilst there is no reason to doubt these claims, it is anything but a reasonable expectation. Something fuzzy in its position can be detected with 10×21 binoculars but, conversely, it can be out of reach on clear-looking nights with 15×70 binoculars. It can sometimes be seen quite clearly as a galaxy with a 9×50 finderscope.

This is certainly the brightest spiral galaxy that can be seen almost face on, as opposed to the almost edge-on appearance of the Andromeda Galaxy (M31). Its position relatively close to M31 in the sky is no co-incidence. The two are members of the Local Group of galaxies, including our own Milky Way. As to how far away it is, modern estimates give 2.9–3 million light years, although figures as low as 2.2 million light years have been quoted. This is due to refinements of the cosmic scale. Tomorrow, someone could make a discovery that reveals it is just in our backyard! What is fairly certain is that it is slightly further away than M31 and (assuming a distance of about 3 million light years) makes it about 60,000 light years across. This is about typical for a spiral galaxy but it makes it smaller than our Milky Way and much smaller than M31.

It has several globular star clusters, although these are more readily visible with research instruments or very large amateur telescopes. It also has several star-forming regions that have been known about since William Herschel's time.

Fig. 4.25 M33 by Mike Deegan

With a high northerly declination, it can appear at the zenith from the southern United States and even at a healthy 60° for northern regions of the UK and Canada. It is a difficult object from 35° south of the equator, where much of the population of the southern hemisphere lives.

Where to Find

M33 can be quite difficult to find, as its brightness is spread out over a large area. Once at a star party from a dark site, we were able to see it very clearly in a 9×50 finderscope but this isn't an everyday occurrence. In fact, from suburban skies, it can even be a challenge for 15×70 binoculars. The best way to find it is draw an imaginary line from Alpha Trianguli to Beta Andromedae. Move a third of the way and then slightly south east. If you can't see M31 clearly, its best to save M33 for another night (Fig. 4.26).

Object Type	Spiral galaxy
Declination	30° 39 min
Right Ascension	1 h 33.9 min
Magnitude	5.7

What It Looks like

Through modest instruments, there is little to suggest the nature of M33. It just looks like a misty patch, which could equally be a nebula or distant open star cluster. Through 15×70 binoculars, it appears like a misty patch and it is sometimes possible to make out the spiral arms. However, on more than one occasion the spiral arms (which appear as smaller fuzzy objects) have appeared detached from the host galaxy! This has happened often enough to know it is not just a once off. Through a Maksutov-Cassegrain, it appears like an extended misty patch, with a bright central core and some outlying patches, which do not quite clearly show as spiral arms. Using a normal eyepiece, it is impossible to fit the entire galaxy into the field of view, although it is possible with a focal reducer. From a dark site, a 127 mm Maksutov (actually belonging to another society member) showed dust lanes but do not expect this to be a normal occurrence.

Charles Messier's Original Notes

25 August 1764
(RA: 01 h 40 m 37 s, 20d 09' 17"; Dec: +29d 32' 25"; Diameter: 0d 15')

Nebula discovered between the head of the northern fish [Pisces] & the Great Triangle, a short distance of a star of 6th magnitude: the nebula has a whitish light, an almost even brightness, however a little more luminous two-thirds of its diameter along, & contains no star. It is hard to see with an ordinary telescope of 1 ft. Its position is determined by comparing with α Trianguli. Reviewed September 27, 1780.

How to View

The best way to see M33 is to use the largest aperture with an eyepiece/accessory combination that can achieve a 2° field of view. An example would be a 130 mm telescope with a focal ratio of F/5 used with a 32 mm focal length Plössl eyepiece. This yields a field of view of about 2.5°. If you have a large budget and can afford exceptional eyepieces, you can use larger aperture instruments with a low focal ratio.

Photographic Details

Figure 4.25 was taken using a 10" F5 Newtonian, Starlight Xpress H×916, guided with a 80ED refractor with a modifed Toucam. It consisted of 25 5 min exposure luminance frames, ten 10 min exposures of hydrogen alpha frames and 2 5 min exposures of RGB binned 2x2 (Fig. 4.27).

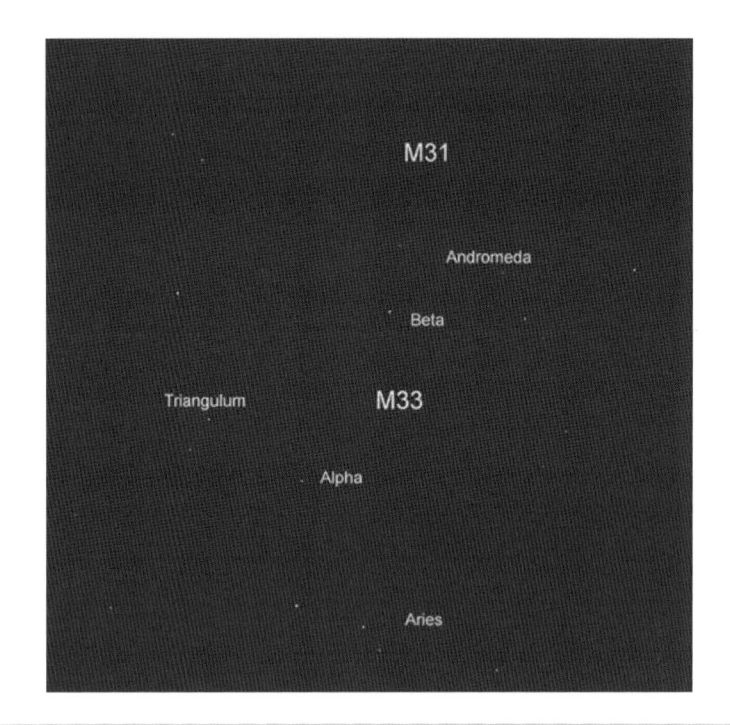

Fig. 4.26 How to find M33

Fig. 4.27 M33 modified

M34

M34 is an open star cluster in Perseus. It is often overlooked in favor of the better-known neighbors, such as the Pleiades and Perseus Double Cluster. Nevertheless, it is a splendid object in its own right and its appearance in large binoculars is reminiscent of the Jewel Box in the Southern Cross, which isn't visible from anywhere more than 30° north of the equator. It has about a 100 members, the brightest being about magnitude 7.9. It is spread over an area of about 35 arcminutes and is about 1,400 light years away. It was known by Giovanni Battista Hodierna before 1654 but would have been visible as a misty patch to the unaided eye before then.

Where to Find

M34 is about 3° WNW of the famous eclipsing binary star Algol. It can be seen easily in binoculars or most finderscopes, so is quite easy to find (Fig. 4.28).

Object Type	Open star cluster
Declination	42° 47 min
Right Ascension	2 h 42.0 min
Magnitude	5.5

Fig. 4.28 How to find M34

What It Looks like

M34 is reminiscent of the Jewel Box through a good pair of binoculars or 80 mm short tube refractor. However, a 127 mm Maksutov-Cassegrain really brings it out and there's a few nice background stars (which aren't actual cluster members) that make its apparent size about a degree and a half. It really is worth the trouble to use a larger aperture instrument on it. Some fainter background stars are lost on poor nights but the overall effect isn't lost. On one viewing it was similar to the views of the Beehive (M44) through binoculars.

Charles Messier's Original Notes

25 August 1764
(RA: 02 h 27 m 27 s, 36d 51' 37"; Dec: +41d 39' 32"; Diameter: 0d 15')

Cluster of small stars, between the head of Medusa & the left foot of Andromeda, almost parallel to γ [Andromedae]: with an ordinary telescope of 3 ft, the stars can be told apart. Its position was determined from β [Persei], the head of Medusa.

How to View

To get in some of the background stars, try to keep the field of view to about 1°. For long focal length instruments, this means using long focal length eyepieces or even a focal reducer, which is recommended.

Photographic Details

This was taken using a 110 mm/F7 refractor and SBIG ST2000×cm camera cooled to 25° below ambient temperature. Fifteen images, each of 1 min duration were taken and stacked and processed using *Nebulosity* (Stark Software) (Fig. 4.29).

Figure 4.30 is a closer representation of what one can really expect from a 127 mm Maksutov. Some of the fainter stars are lost but not too many.

M34 is a tough object with only a compact digital camera. Attempts to snap it with an 80 mm refractor failed but its northerly declination made it possible to run an 8 s exposure using my Maksutov with 24× magnification and a light pollution reduction filter (Fig. 4.31).

Well it is a nice object, so we decided to include a snap with a wider field of view (Fig. 4.32).

Fig. 4.29 M34 photographed by Anthony Glover

Fig. 4.30 Modified M34

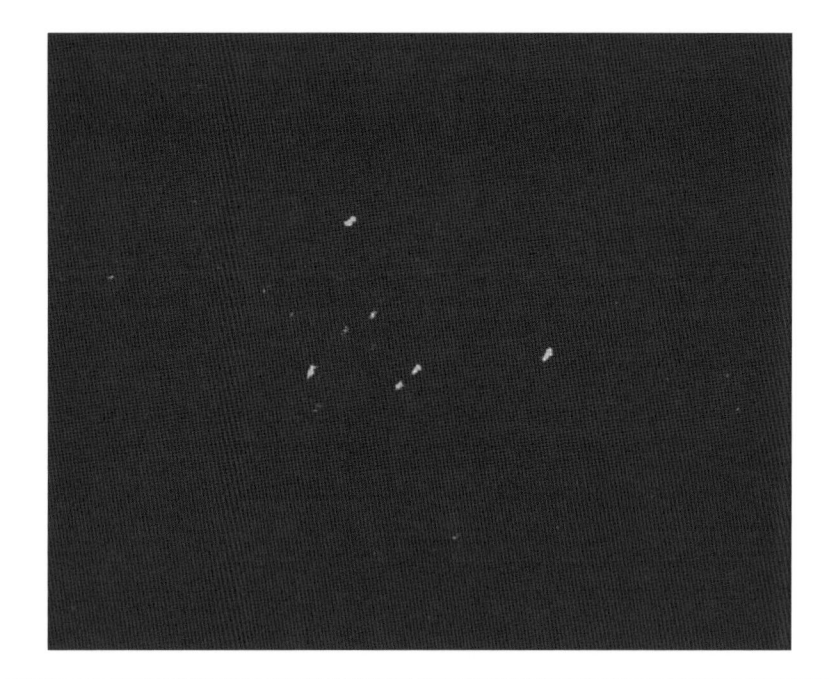

Fig. 4.31 M34 by Philip Pugh

Fig. 4.32 M34 by Anthony Glover

M35

M35 is an open star cluster in Gemini and is quite well-known. It can be visible to the unaided eye, so although its "discovery" is credited to Philippe Loys de Chéseaux, it may have been known about before then. It is about 2,800 light years away. It contains hundreds of stars in an almost circular pattern of about 28 arcminutes. There are over 100 stars brighter than 13th magnitude and some stars resolve in quite modest instruments. Inexperienced observers may confuse it with nearby NGC2158, which is smaller and fainter but more compact. As this area of sky is rich in background Milky Way stars, M35 can sometimes get lost against the background on very clear nights.

Where to Find

M35 is found in Gemini near the borders with both Orion and Taurus. Although there are no particularly bright stars nearby, it is bright enough to be easy to find (Fig. 4.34).

Fig. 4.33 M35

Object Type	Open star cluster
Declination	24° 20 min
Right Ascension	06 h 08.09 min
Magnitude	5.3

What It Looks like

The best way to describe M35 is to say that it looks like how a rich open star cluster should look like. It looks like what it should do: a collection of stars of various luminosities together in the same field of view. It is an easy enough object in large binoculars but a larger aperture instrument really brings it to life and is breathtaking. Like many objects in the same region of sky, the best views include the background stars.

Charles Messier's Original Notes

30 August 1764
(RA: 05 h 54 m 41 s, 88d 40' 09"; Dec: +24d 33' 30"; Diameter: 0d 20')

Cluster of very small stars, near the left foot of Castor, near the stars μ and η of this constellation [Gemini]. M. Messier reported its position on the chart of the Comet of 1770. *Memoirs of the Academy for the year 1771, plate VII*. Reported in the *English Atlas*.

How to View

Although the official size of M35 is "only" 28 arcminutes, a field of view of about a degree and a half is needed to take in the excellent Milky Way background. The only other tip is to use as large an aperture as you can, as many stars are 13th magnitude. A 127 mm Maksutov-Cassegrain does it proud. Its overall brightness makes it just too faint to make my Usual Suspects list, as it is a tough object for 50 mm binoculars.

Photographic Details

This was taken using a 110 mm/F7 refractor and SBIG ST2000xcm camera cooled to 25° below ambient temperature. Fifteen images, each of 1 min duration were taken and stacked and processed using *Nebulosity* (Stark Software) (Fig. 4.33).

The view of M35 through the Maksutov is very similar to this photograph. It is a superb object.

Fig. 4.34 How to find M35

M36

M36 is an open star cluster in Auriga. It is 12 arcminutes across and most estimates of its distance agree at 4,100 light years. It was discovered by Giovanni Battista Hodierna some time before 1654. It has about 60 members, with the brightest one being about 6th magnitude. It appears as a misty patch in 15x70 binoculars. Although it is circumpolar from northern parts of the UK and Canada, it is best seen in winter when it is high enough to avoid extinction.

Where to Find

M36 forms a close group with M37 and M38, with M37 being marginally the brightest of the three. None of them is easy to locate against the stellar background but are certainly bright enough to see in a decent finderscope, provided that Auriga is well above the horizon. It is a good idea to familiarize yourself with their positions through binoculars before attempting telescopic observations (Fig. 4.36).

Object Type	Open star cluster
Declination	34° 08 min
Right Ascension	05 h 36.1 min
Magnitude	6.3

What It Looks like

M36 isn't easy enough to make the Usual Suspects List. However, it can be seen as a misty patch through 15×70 binoculars on a good (but not typical) night. It can be found in a 9×50 finderscope as well. As you progress through higher apertures, some of the stars are resolved and this gives it a particular beauty. However, M37 is similar but more impressive.

Fig. 4.35 M36 by Anthony Glover

Fig. 4.36 How to find M36

Charles Messier's Original Notes

2 September 1764
(RA: 05 h 20 m 47 s, 80d 11' 42"; Dec: +34d 08' 06"; Diameter: 0d 09')

Star clusters in the chariot [Auriga], near the star φ: with an ordinary telescope of 3 ft and a half we have a hard time seeing the stars, the cluster contains no nebulosity. Its position is determined from φ Aurigae.

How to View

M36 is best viewed when some but not all of its stars are resolved. Its relatively small size of 12 arcminutes allows higher magnification to resolve the gaps between the stars. Although it is a binocular object on a good night, it is best to use a telescope, preferably with an aperture of at least 100 mm.

Photographic Details

This is a good representation of how M36 appears in a 127 mm Maksutov, so needs no modification.

Figure 4.35 was taken using an 80 mm F/6 APO and SBIG ST2000×CM CCD. Twenty 1 min exposures were taken. Detail: Starlight Xpress SXV H9 and WO 80 mm F6 refractor. Twenty 1 min subs IR blocking filter only.

M37

M37 is an open star cluster in Auriga. It is 24 arcminutes across. Although its combined magnitude is slightly brighter than nearby M36, the brightness is spread out over a much larger area, so is more difficult to spot. It is also slightly further south than its neighbor, so is more prone to extinction when near the horizon. It was also discovered by Giovanni Battista Hodierna before 1654. It contains 150 stars brighter than magnitude 12.5 and about 500 overall. It appears as a misty patch in 15×70 binoculars. There are several estimates of its distance between 3,600 and 4,400 light years.

Where to Find

M37 is in the same part of the sky as M36 and M38, so please refer to the diagram for M36.

Object Type	Open star cluster
Declination	32° 33 min
Right Ascension	05 h 52.4 min
Magnitude	6.2

What It Looks like

M37 isn't easy enough to make the Usual Suspects List. However, it can be seen as a misty patch through 15×70 binoculars on a good (but not typical) night. As you progress through higher apertures, some of the stars are resolved and this gives it a particular beauty. Its irregular shape is obvious and if you didn't know what it was, you would say it's an irregular galaxy with a few foreground stars, possibly forming a cluster. It is well worth a look, as it looks breathtaking through a 127 mm Maksutov on a clear night, as some of its stars become resolved.

Fig. 4.37 M37 by Anthony Glover

Charles Messier's Original Notes

2 September 1764
(RA: 05 h 37 m 01 s, 84d 15′ 12″; Dec: +32d 11′ 51″; Diameter: 0d 09′)

Cluster of small stars, not far from the previous one, as above the parallel of χ Aurigae; the stars are smaller and closer and contain nebulosity, with an ordinary telescope of 3 ft & a half, it is difficult to see stars: this cluster is reported on the chart of the second Comet of 1771, *Memoirs of the Academy for the year 1777*.

How to View

M37 is best viewed when some but not all of its stars are resolved. Its relatively large size of 24 arcminutes allows moderate magnification only, say up to 80× to get the whole cluster into the field of view. Although it is a binocular object on a good night, it is best to use a telescope, preferably with an aperture of at least 100 mm.

Photographic Details

This is a good representation of how M37 appears in a 127 mm Maksutov, so needs no modification.

Figure 4.37 was taken using an 80 mm F/6 APO and SBIG ST2000×CM CCD. Twenty 1 min exposures were taken.

M38

M38 is the faintest of the three main open star clusters in Auriga. It is also 21 arc-minutes in extent, so may be missed when M36 is visible. It was also discovered by Giovanni Battista Hodierna before 1654. Its brightest star is magnitude 7.9. Like its neighbors, it appears as a misty patch to binoculars. With an estimated distance of about 4,200 light years, it is likely to be near M36 and M37.

Where to Find

M38 is in the same part of the sky as M36 and M37, so please refer to the diagram for M36.

Object Type	Open star cluster
Declination	35° 50 min
Right Ascension	05 h 28.4 min
Magnitude	7.4

What It Looks like

After viewing M37 and M36, M38 is somewhat of an anti-climax. It looks misty through binoculars but a medium aperture telescope resolves many of its stars and they form a rather sparse pattern. It is more of an object to cross off than one of the highlights of the Messier Catalogue.

Charles Messier's Original Notes

25 September 1764
(RA: 05 h 12 m 41 s, 78d 10′ 12″; Dec: +36d 11′ 51″; Diameter: 0d 15′)

Cluster of small stars in the chariot [Auriga], near the star σ, not far from both previous clusters, which is contained in a square shape & contains no nebulosity, if we care to examine it with a good telescope. The area subtends 15 min of arc.

How to View

A large aperture is helpful to capture as much light as possible but its best to keep the field of view as large as possible to keep the distances between the stars small to overcome the sparse effect. With a size of 21 arcminutes and a nice background field, you need to aim for field of view of about one degree.

Photographic Details

This is a good representation of how M38 appears in a 127 mm Maksutov, so needs no modification.

Figure 4.38 was taken using an 80 mm F/6 APO and SBIG ST2000×CM CCD. Twenty-one 1 min exposures were taken.

Fig. 4.38 M38 by Anthony Glover

M39

M39 is an open star cluster in Cygnus. It is quite sparse with about 30 members spread out over 32 arcminutes. Its integrated brightness is a matter of doubt and, indeed is difficult to see without its individual stars, the brightest being magnitude 6.83. It can be surprisingly difficult on very clear nights, as it can get lost in the Milky Way background. Although it is listed as one of Charles Messier's original discoveries, some sources claim it was known by Guilliame Le Gentil and even Aristotle. It is said to be visible to the unaided eye from good sites, so this is possible. It is about 800 light years away. It is circumpolar from the United Kingdom and northern parts of North America.

Where to Find

M39 is a hard object to find, being in north eastern Cygnus, about 9° east and 4° north of Deneb. There are not any nearby stars brighter than 4th magnitude, although it is in the plane of the Milky Way (Fig. 4.40).

Fig. 4.39 M39 (Courtesy of Michael Oates, Manchester Astronomical Society)

Object Type	Open star cluster
Declination	48° 26 min
Right Ascension	21 h 32.2 min
Magnitude	4.6

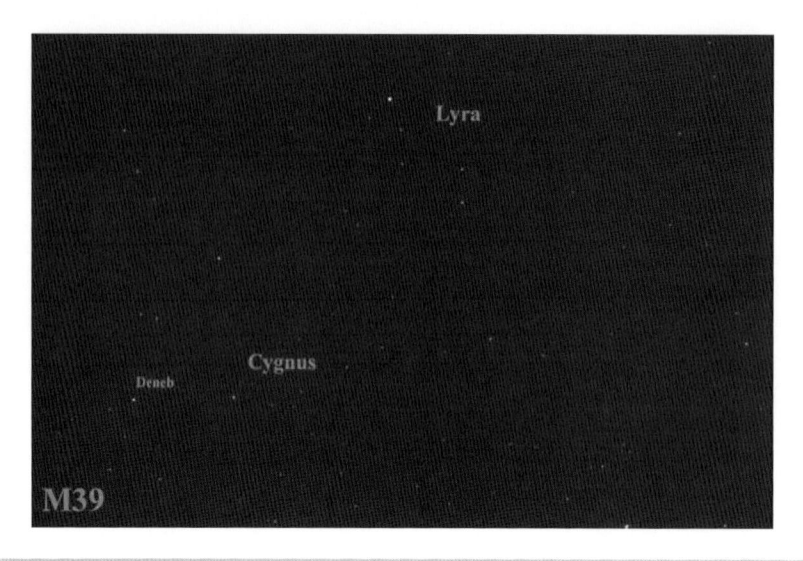

Fig. 4.40 How to find M39

What It Looks like

It can often be missed through binoculars and it is often easier to pot on slightly cloudy nights when the background stars are dimmed. Through a Maksutov, a group of brighter stars stand out from the background, with no obvious shape. As for the fainter stars, it is not obvious which are true cluster members and which are background.

Charles Messier's Original Notes

24 October 1764
(RA: 21 h 23 m 49 s, 320d 57' 10"; Dec: +47d 25' 00"; Diameter: 1d 00')

Cluster of stars near the tail of the swan [Cygnus]; we can see with a regular telescope of 3 ft & a half.

How to View

On really clear nights, aperture can work against you, as it brings more Milky Way background stars into view. It is best to wait for a slightly hazy night or use a small aperture.

Photographic Details

Figure 4.39 is also a good representation of how M39 appears in a 127 mm Maksutov.

M40

M40 is a double star and Charles Messier found it in 1764 while searching for a nebula that was recorded (erroneously) at this position. The components are magnitudes 9.0 and 9.3 and are separated by about 50 arcseconds, making them visible in many small aperture instruments. Their distance is about 510 light years, although evidence seems to suggest that they are not physically associated. M40 is circumpolar from the United Kingdom and most parts of the USA.

Where to Find

M40 is about 2° north east of Delta Ursae Majoris in the Plough asterism. It is bright enough to be visible in a 70 mm binoculars (Fig. 4.42).

Object Type	Double star
Declination	58° 05 min
Right Ascension	12 h 22.4 min
Magnitude	8.4

Fig. 4.41 M40 (NASA/courtesy of nasaimages.org)

What It Looks like

Through 70 mm binoculars, M40 looks quite unremarkable but there are enough background stars visible to suggest that there's some nebulosity. This was a surprise because it is not in a particularly rich part of sky. The view through a 127 mm Maksutov shows the double star split very well and the number of background stars definitely suggests that they could be mistaken for something fuzzy in smaller instruments.

Charles Messier's Original Notes

24 October 1764
(RA: 12 h 11 m 02 s, 182d 45' 30"; Dec: +59d 23' 50"; Diameter: –)

Two stars very close to each other & very small, placed at the rise of the tail of the Great Bear [Ursa Major]: it has been difficult to distinguish them with a regular telescope of 6 ft. In seeking the nebula placed over the back of the Great Bear, reported in the book of the *Figures of the Stars*, for 1660, 183d 32' 41" of right ascension & 60d 20' 33" northern declination, that M. Messier could not see, he has observed these two stars.

How to View

Fortunately, although M40 is not an exciting object, it is not a demanding one either. As a wide double star, it should be visible in 50 mm binoculars.

Fig. 4.42 How to find M40

Photographic

This is a good representation of how M40 appears in a 127 mm Maksutov, so needs no modification (Fig. 4.41).

M41

M41 is an open star cluster in Canis Major. It occupies an area 38 arcminutes across but has about 100 stars. This does not include the 6th magnitude star 12 Canis Majoris, which is about half the distance to M41 (2,300 light years). It was discovered by Giovanni Battista Hodierna before 1654 but may have been known previously to Aristotle. It fails to make the Usual Suspects List because many of its stars are lost to extinction from England. From Australia, New Zealand and Chile, it is a spectacular sight, somewhat fainter than the Pleiades (M45) but brighter than the Beehive (M44).

Where to Find

M41 can be found about 5° south of Sirius. Were it not for its southerly declination, it would be visible to the unaided eye from the northern hemisphere. From 35° south of the equator, it is (Fig. 4.44).

Object Type	Open star cluster
Declination	−20° 44 min
Right Ascension	06 h 46.0 min
Magnitude	4.5

What It Looks like

M41 looks quite superb when viewed near the zenith from sub-tropical locations in the southern hemisphere, even with modest instruments. However, it is a different proposition when viewed from northern Europe and Canada, attaining an altitude of about 20° at best. In binoculars, it resembles the Jewel Box at first sight. In a 127 mm Maksutov, some of the fainter stars become visible, especially on a clear night, although nowhere near the 100 members are visible at an elevation of 20° or more. On a clearer night, some of the fainter stars twinkle in and out of visibility, giving a rather pleasant effect.

Charles Messier's Original Notes

16 January 1765
(RA: 06 h 35 m 53 s, 98d 58' 12"; Dec: −20d 33' 00"; Diameter: –)

Fig. 4.43 M41 (NASA/courtesy of nasaimages.org)

Star cluster below Sirius, near ρ Canis Majoris; this cluster has a nebulous appearance in an ordinary telescope of a foot: This is a small cluster of stars.

How to View

If you live in the southern hemisphere, it is near the zenith in summer and an easy target for just about any instrument. From suburban locations of about 50° latitude north of the equator, it needs something a bit more substantial, such as 70 mm aperture binoculars. However, potentially the best view would be through a large reflector with a wide field of view.

Photographic Details

Figures 4.43 and 4.45

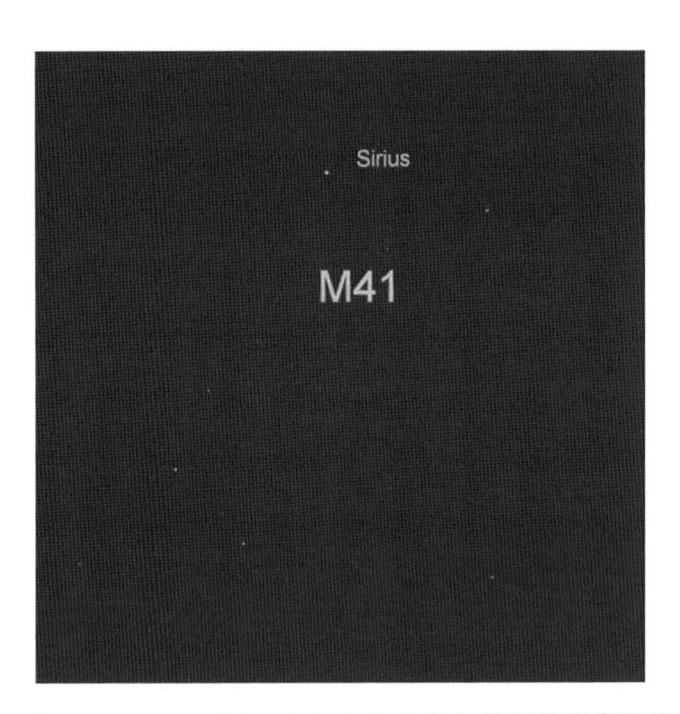

Fig. 4.44 How to find M41

Fig. 4.45 M41 modified (NASA/courtesy of nasaimages.org)

M42

This is also known as the Great Orion Nebula. It also contains an "extension" known as M43 that was once thought to be a separate object. It is a vast star-forming region about 1,270 light years away (according to very recent estimates) and many of the stars that it has borne can be seen. It was probably discovered 1610 by Nicholas-Claude Fabri de Peiresc. It measures 85×60 arcminutes across.

Where to Find

Figure 4.47

Object Type	Nebula
Declination	−5° 23 min
Right Ascension	5 h 35.4 min
Magnitude	4

Fig. 4.46 M42

How to Find

M42 is not far to the south of Orion's belt and can normally be seen as a fuzzy patch with the unaided eye.

What It Looks like

M42 can be seen as a misty patch to the unaided eye on a clear night, even from suburban locations. It can even be seen in the "How to Find" photograph of Orion and its surrounding area. Even small binoculars will show it as an extended misty patch and larger binoculars will show four stars that form a rhombus shape and are known as the Trapezium.

As you progress through larger aperture instruments, you can see more and more detail. On an exceptional night, you can see the butterfly shape of the nebula quite clearly and some of the light and dark regions become resolved in a 127 mm Maksutov. On a more typical cloudless night, you will need a 200 mm aperture instrument to capture the outer regions of the nebula but the "butterfly" shape of the central region is quite apparent in less than perfect conditions.

Charles Messier's Original Notes

4 March 1769
(RA: 05 h 23 m 59 s, 80d 59' 40"; Dec: −05d 34' 06"; Diameter: –)

Position of the beautiful nebula in the sword of Orion, the star θ contained therein with three other smaller stars that we can see with good instruments. Messier went into great detail as to this Great Nebula; he gave a drawing made with the highest care, we can see it in the *Memoirs of* the *Academy, year 1771, plate VIII*. It was Huygens who discovered it in 1656: it has been observed since by many astronomers as reported in the *English Atlas*.

How to View

Like many of the larger Messier objects, the large size of M42 means that the battle between aperture and field of view takes place. Ideally, you need a 2° field of view to show the whole extent properly, which limits the aperture to about 150 mm, unless you can use eyepieces with a very wide field of view. Focal reducers are of great help here.

Fig. 4.47 M42 by Anthony Glover

Photographic Details

Figure 4.46 was taken using an 80 mm refractor at 6× magnification using afocal projection, a light pollution reduction filter and compact digital camera with 8 s exposure. It is quite a good approximation of M42 as seen through many binoculars.

Figure 4.48 shows M42. It was taken using a 80 mm/F6 refractor and Starlight Xpress SXVH9 camera cooled to 25° below ambient temperature. Four images, each of 3 min duration were taken (each for Luminosity, Red, Green and Blue channels) and stacked and processed using *Nebulosity* (Stark Software). Figure 4.49 shows M42 as it looks through a 127 mm Maksutov.

M43

M43 is a nebula that appears as a separate object to the Great Orion Nebula (M42). In fact it isn't, it's just that the "join" is obscured by a dark dust lane. It is the same distance as M42 and is 20×15 arcminutes in extent. It was discovered by Jean-Jacques d'Ortous de Mairan before 1731.

Where to Find

M43 is right next to M42 and would otherwise be difficult to find. Please refer to Fig. 4.46 for the position of M42.

M43

Object Type	Nebula
Declination	−5° 16 min
Right Ascension	5 h 35.6 min
Magnitude	9

What It Looks like

M43 is the "extension" to M42 that appears a bit like a wizard's hat. It is much fainter than M42 and can be missed in small instruments.

Charles Messier's Original Notes

4 March 1769
(RA: 05 h 24 m 12 s, 81d 03′ 00″; Dec: −05d 26′ 37″; Diameter: –)

Position of the small star which is surrounded by nebulosity and which is below the nebula of the sword of Orion. M. Messier has reported it in the drawing of the Great [Nebula?].

Fig. 4.48 How to find M42

Fig. 4.49 M42 modified

Fig. 4.50 Simple shot of M42

How to View

Due to its large size and dimness, it needs at least a 100 mm aperture instrument on a clear night. It is also imperative that Orion is clear of the horizon and you can see some structure in M42. If M42 looks disappointing, it is not worth trying for M43.

Photographic Details

Please refer to the section on M42.

M44

M44 is more commonly known as the Beehive or Praesepe (the manger). It was known to Aratos in 260 BC and can be seen with the unaided eye on a clear night. It is 95 arcminutes across, so a fine object for binoculars. It is about 577 light years away and has 200 known members, although the area contains 350 stars, some of which are foreground, others background and many "don't knows". It is one of my personal favourites and before I took astronomical photographs, did a drawing of it, by observing it over several nights with my 15×70 binoculars (Fig. 4.53).

Fig. 4.51 M44 by Anthony Glover

Where to Find

The main difficulty in finding M44 is that Cancer (its host constellation) is very faint and quite difficult to see from suburban skies. The good news is that you can find it quite easily if Pollux and Regulus are both above the horizon. On a line joining Pollux to Regulus, M44 is about 2/5 of the way along and can be seen easily in binoculars or a finderscope (Fig. 4.52).

Object Type	Open star cluster
Declination	19° 41 min
Right Ascension	8 h 40.1 min
Magnitude	3.7

What It Looks like

M44 can be seen as a misty patch to the unaided eye on a clear night, from suburban locations. Even on a poor night, the main asterism of 6th and 7th magnitude stars forms a pattern reminiscent of a swarm of bees in binoculars. It used to be one of my favorites in my 76 mm catadioptric when I used it regularly. As you progress through the apertures,

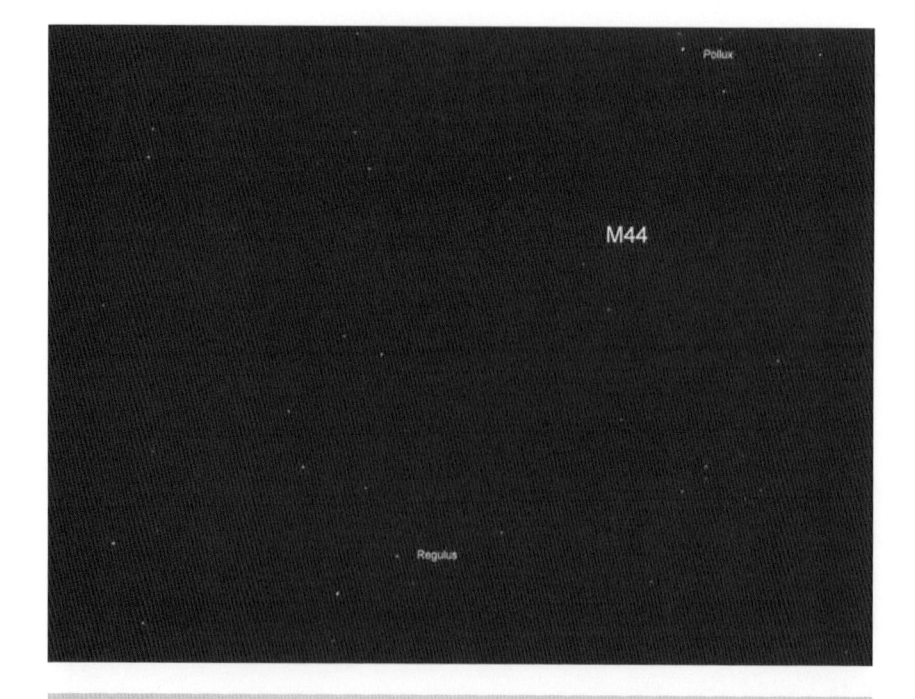

Fig. 4.52 How to find M44

more and more stars become visible. The best ever view from personal experience (unsurprisingly) was with my 127 mm Maksutov-Cassegrain, although some of the outlying members could not fit into the field of view. It also suggests that the "official" figure for the angular size could be more like 2.5°. Some of the brighter members of the cluster resemble a series of double and triple stars that happen to be in the same field of view, suggesting that at least some might be gravitationally attached true binaries.

Charles Messier's Original Notes

4 March 1769
(RA: 08 h 07 m 22 s, 126d 50' 30"; Dec: +20d 31' 38"; Diameter: –)

Star cluster known by the name of nebula of Cancer, the position reported is that of the star C.

How to View

M44 can be enjoyed in binoculars and is one of the recommended objects for casual viewing. A Dobsonian reflector of short focal length is potentially the best way to see it. The key to viewing it is to ensure that the field of view should be at least a bit more than its angular size. 2° fits it in nicely, whereas the maximum field of view achievable with my Maksutov-Cassegrain and current accessories is about 1.7°, just squeezing it in. It also looks very pleasing to the eye in a Startravel 80 short tube refractor.

Photographic Details

M44 is a great photographic subject for modest equipment. Figure 4.53 shows a simple example.

Figure 4.53 shows M44 taken using a compact digital camera using afocal projection with a Skywatcher Startravel 80 refractor at 6× magnification with a light pollution reduction filter. The exposure time was 8 s.

Figure 4.54 shows a lot more detail and is very similar to the view through a 127 mm Maksutov. Many fainter stars are visible. It was taken using a 110 mm/F7 refractor and SBIG ST2000×cm camera cooled to 25° below ambient temperature. Fifteen images, each of 1 min duration were taken and stacked and processed using *Nebulosity* (Stark Software).

Fig. 4.53 M44 Drawing

Fig. 4.54 M44 by Philip Pugh

M45

M45 is more commonly known as the Pleiades or Seven Sisters. It was known about by Homer in 750 BC but has probably been known since the dawn of civilisation, as it is a prominent object easily visible to the unaided eye. It is an open star cluster in Taurus and has an apparent dimension of nearly 2° across, although research instruments have detected brown dwarves in the cluster, suggesting that it is nearly 5° across. Various estimates of its distance range from 380 light years to 440. There is also some nebulosity associated with M45, yet it is not the remains of the molecular cloud that gave birth to the cluster, merely another object that is "just passing by".

M45 is probably the easiest object in the Messier Catalogue to view. It looks spectacular even in very small binoculars and in larger ones in the 70–80 mm class, you really get the "wow" factor. It is said that you can see 103 members with a pair of 110 mm binoculars and about 80 can be seen on a clear night with 70 mm binoculars. A well known challenge is to count how many you can see without optical aid. The record is 19 and my personal best is 12. Most people can see six or eight.

Where to Find

Figure 4.57

Fig. 4.55 M45 by Anthony Glover

Object Type	Open star cluster
Declination	24° 07 min
Right Ascension	3 h 47.0 min
Magnitude	1.6

How to find

M45 is to the north west of Aldebaran, the brightest star in Taurus and can also be found under the right (east) foot of Perseus. It is visible to the unaided eye most nights.

What It Looks like

M45 can be seen as tight grouping of stars to the unaided eye on a cloudless night, even from suburban locations. It can even be seen in the "How to Find" photograph, above. The main asterism is somewhat reminiscent of the Big Dipper/Plough asterism in Ursa Major. As aperture increases and/or conditions improve, a small chain of stars traversing from north to south of the cluster appears. In larger instruments, the number of stars visible increases and under exceptionally good conditions it is possible to see the nebulosity often seen in photographs.

Charles Messier's Original Notes

4 March 1769
(RA: 03 h 33 m 48 s, 53d 27' 04"; Dec: +23d 22' 41"; Diameter: –)

Star cluster, known under the name of the Pleiades. The position reported is of the star Alcyone.

How to View

Unfortunately, this is where a 127 mm Maksutov-Cassegrain falls short. Pointing it at M45 certainly shows lots of stars but the narrow field of view makes it impossible to see the whole cluster at once. The best you can realistically achieve is to get the main asterism into the field of view. It really is better seen in binoculars and short tube refractors. Apart from getting the whole of it into the field of view, there is not really anything else you can do to improve the appearance.

M45 165

Photographic Details

Figure 4.56 shows about 25 of the brightest stars of M45 and was taken using the same techniques as for M44.

Figure 4.55 was taken using 6 × 5 min exposures with an 80 mm/F6 ED refractor, SBIG St2000×cm, dark subtracted, stacked and processed in *Nebulosity*. It looks very similar to the view through a 127 mm Maksutov, except that a Maksutov does not show diffraction spikes. The nebulosity is only visible on a good night from a dark site or using a light pollution reduction filter (Fig. 4.55).

Fig. 4.56 Simple shot of M45

Fig. 4.57 How to find M45

Chapter 5

M46–M68

M46

M46 is an open star cluster in Puppis (part of the original constellation of Argo Navis). It was discovered by Charles Messier in 1771. It is 27 arcminutes across and contains about 150 stars, with the brightest being about magnitude 8.7. It is 5,400 light years away. It is quite often overlooked for the more famous M41 in Canis Major. It is never at its best when seen from the northern hemisphere but it looks quite superb 35° south of the equator, where most of the southern hemisphere's population is gathered.

Where to Find

Although M46 and M47 are both in the southern constellation of Puppis, which is largely hidden from us in the UK, they are surprisingly far north, in fact both are further North than Sirius. They are very close and should be seen in the same binocular field of view. There are no bright stars nearby, the brightest being Alpha Monocerotis, about 5° north of M46. M47 is about a degree and a half west and slightly north of M46. You will probably notice it first, as it is about 1.5 magnitudes brighter. When scanning using a finderscope or binoculars, it is best to start with Sirius scan about an hour of right ascension east and a degree and half north (Fig. 5.2).

Fig. 5.1 M46 (NASA/courtesy of nasaimages.org)

Object Type	Open star cluster
Declination	−14° 49 min
Right Ascension	7 h 41.8 min
Magnitude	6

What It Looks Like

M46 can be seen as a misty patch in binoculars but has no obvious bright stars. In a 127 mm Maksutov, it comes alive, with many stars flickering in and out of visibility, due to the twinkling effect, as it never gets high in the sky from northern Europe and Canada. It is a much richer cluster than the nearby M47.

Charles Messier's Original Notes

19 February 1771
(RA: 07 h 31 m 11 s, 112d 47′ 43″; Dec: −14d 19′ 07″; Diameter: –)

Clusters of very small stars between the head of the Great Dog [Canis Majoris] and the two hind legs of the Unicorn [Monoceros], found by comparing [it] to the star two Navis [a star now in Puppis] of 6th magnitude, according to Flamsteed; we cannot see these stars unless one has a good telescope, the cluster contains a few nebulas [nebulosity].

How to View

Due to its large size of 27 arcminutes, it needs a wide field of view to get it in, so will not tolerate high magnification if you want to see the whole cluster. A 120 mm instrument with a focal length of about 600 mm should be about right, although larger apertures will be needed when it is low in the sky but you will lose some of the outlying members.

Photographic Details

Figures 5.1 and 5.3

Fig. 5.2 How to find M46

Fig. 5.3 M46 modified (NASA/courtesy of nasaimages.org)

M47

Like M46, M47 is an open star cluster in Puppis. It was discovered before 1654 by Giovanni Battista Hodierna, although Charles Messier was unaware of this discovery. It is closer than M46, at 1,600 light years and has fewer stars (50) spread out over a wider area (30 arcminutes). However, the stars are brighter, with the brightest being of magnitude 5.7. Messier made a mistake in noting the position of this cluster and it was found later by William and Caroline Herschel (amongst others).

Where to Find

As M47 is very close to M46, please refer to the *How to Find* section for M46. In fact, M47 is seen first as it has some bright stars.

Object Type	Open star cluster
Declination	−14° 30 min
Right Ascension	7 h 36.6 min
Magnitude	5.0

What It Looks Like

M47 makes an interesting contrast to M46. To binoculars, it appears like a sparse grouping of 5th/6th magnitude stars but to a 127 mm Maksutov, its true nature is revealed with many background stars twinkling in and out of view. It is not as rich in fainter stars as M46 but is still a nice sight in a medium aperture instrument, while it can be disappointing in binoculars.

Charles Messier's Original Notes

19 February 1771
(RA: 07 h 44 m 16 s, 116d 03' 58"; Dec: −14d 50' 08"; Diameter: −)

Star cluster a little way from the preceding. The stars are brighter; the middle of the cluster was compared to the same star, two Navis. The cluster contains no nebulosity.

How to View

Due to its large size of 30 arcminutes, it needs a wide field of view to get it in, so will not tolerate high magnification if you want to see the whole cluster. A 120 mm instrument with a focal length of about 600 mm should be about right, although larger apertures will be needed when it is low in the sky but you will lose some of the outlying members. Binoculars will reveal the brighter members, which form a tight asterism.

Photographic Details

This is a good representation of how M47 appears in a 127 mm Maksutov, so needs no modification.

M47 was taken using 6×5 min exposures with an 110 mm/F7 APO refractor, SBIG ST2000×cm cooled to −25°, stacked in *Nebulosity* (Fig. 5.4).

Fig. 5.4 M47 by Anthony Glover

M48

M48 is an open star cluster near the head of Hydra. Charles Messier discovered it in 1771 but, unfortunately, made an error in recording its position. Oswald Thomas cleared up the confusion in 1934 but it had also been discovered independently by Caroline Herschel and Johann Elert Bode.

It is 1,500 light years away and has an angular size of 54 arcminutes. There are 50 stars brighter than 13th magnitude, with at least 80 known members in total. The brightest star is of magnitude 8.8, which puts it in the range of 50 mm binoculars.

Where to Find

Despite being fairly bright, M48 is quite difficult to find in a telescope. You can get a rough idea by scanning from Procyon to Alphard (Alpha Hydrae). M48 is about 40% of the way but somewhat south. Whilst most of us think of Hydra as starting with the head, it actually extends south and west of it, so M48 isn't too far from M46 and M47 (Fig. 5.6).

Fig. 5.5 M48 by Anthony Glover

Object Type	Open star cluster
Declination	−5° 48 min
Right Ascension	8 h 13.8 min
Magnitude	5.5

What It Looks Like

M48 is theoretically visible to the unaided eye from a dark site but its size means that it is very unlikely. Through 70 mm binoculars, it appears as a very sparse cluster and more like one of those objects that one crosses off the list to say that one's seen it. It is hard to see how such a sparse cluster could be mistaken for a comet. Through my Maksutov, more faint members can be seen, suggesting a tighter cluster than the binocular view would suggest. However, some individual stars can be resolved.

Charles Messier's Original Notes

19 February 1771
(RA: 08 h 02 m 24 s, 120d 36′ 00″; Dec: −01d 16′ 42″; Diameter: −)

Clusters of very small stars, without nebulosity, this cluster is a short distance from the three stars that are at the start of the tail of the Unicorn [Monoceros].

How to View

Whilst the binocular views are somewhat disappointing, moving up to a higher aperture improves the view but the large angular size means that it is difficult to get the whole object into the field of view with large apertures. A field of view of at least 90 arcminutes is recommended and if you can do this with a 200 mm telescope, this would be best.

Photographic Details

This photograph of M48 was taken using an SBIG ST2000×CM cooled to − 25° with an 80 mm APO Refractor and Astronomik LPR filter. Thirty one minute subframes were stacked.

The view through a 127 mm Maksutov looks very similar (Fig. 5.5).

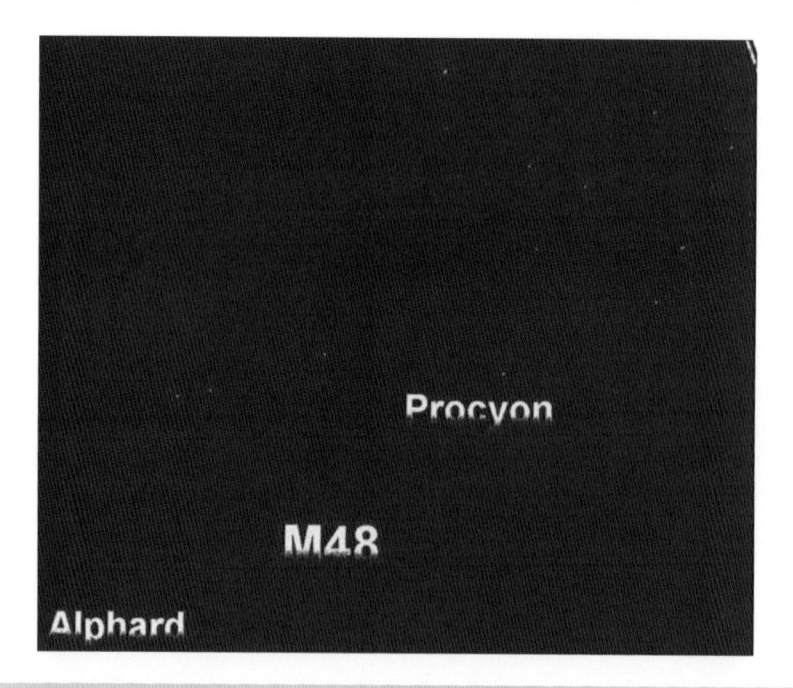

Fig. 5.6 How to find M48

M49

M49 is a giant elliptical galaxy in Virgo. It was discovered by Charles Messier in 1771 but, at the time, its true nature was not known. Although its apparent size is 9×7.5 arcminutes, it a massive 60 million light years away. This suggests a major axis of 160,000 light years, making it much larger than anything in our Local Group. It is a very difficult object for small instruments and, honestly, it can be difficult to distinguish members of the Virgo Supercluster, as most of them appear as fuzzy patches and larger instruments are needed to bring out their shape.

Where to Find

The bad news is that M49 is in a rather sparse patch of sky with no obvious guideposts. Omicron Virginis is the nearest naked eye star and M49 lies about 5° east and slightly south. The good news is that M49 is one of the brighter members of the Virgo Supercluster, competing with M87 for the "title." (Fig. 5.8).

Fig. 5.7 M49 (NASA/courtesy of nasaimages.org)

Object Type	Elliptical galaxy
Declination	8° 00 min
Right Ascension	12 h 29.8 min
Magnitude	8.4

What It Looks Like

M49 is one of the larger and brighter members of the Virgo Supercluster. It appears as a fuzzy patch in a 127 mm Maksutov, with a pronounced nucleus. Only its position differentiates it from M87.

Charles Messier's Original Notes

19 February 1771
(RA: 12 h 17 m 48 s, 184d 26' 58"; Dec: +09d 16' 09"; Diameter: –)

Nebula discovered near the star 'ρ' of the Virgin [i.e. ρ Virginis]. It isn't seen without difficulty with an ordinary telescope of 3 ft & a half. The Comet of 1779 was compared by M. Messier with this nebula on 22 and 23 April: the Comet & the nebula have the same light. M. Messier has reported this nebula on the chart of the Comet, which appears in the volume of the Academy of the same year 1779. Revisited 10 April 1781.

How to View

M49 really needs a large aperture and medium magnification (100x), otherwise it will just appear as a fuzzy patch with a nucleus. A 127 mm Maksutov is simply too small in aperture to make the most of it.

Photographic Details

Figures 5.7 and 5.9

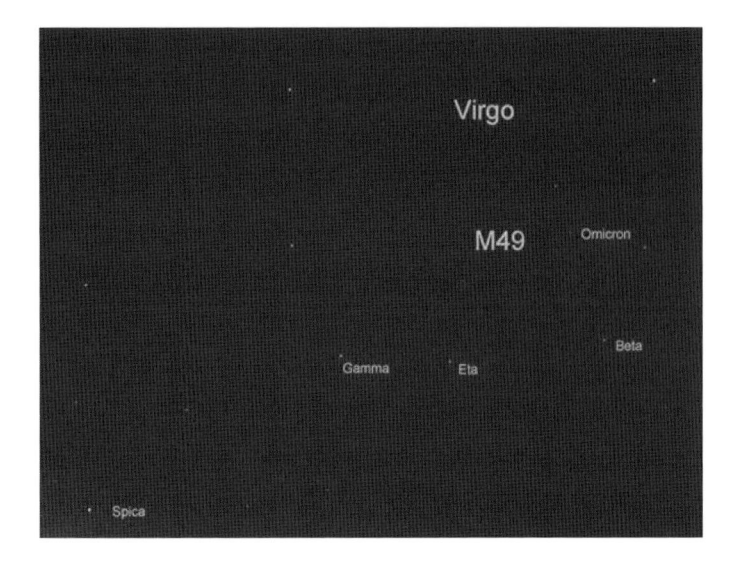

Fig. 5.8 How to find M49

Fig. 5.9 M49 modified (NASA/courtesy of nasaimages.org)

M50

M50 is an open star cluster in Monoceros. It was discovered by Giovanni Cassini in 1711 or before, although Charles Messier was unaware of this previous discovery when he found it in 1772. It is 16 arcminutes across but only the central 10 arcminutes are immediately obvious when using small instruments. It is about 3,200 light years away and has about 200 members, the brightest being 9th magnitude.

Where to Find

M50 is in the faint constellation of Monoceros, which isn't easily visible from suburban skies. The easiest way to find it is to imagine a line joining Sirius to Procyon then it is about a third of the way from Sirius (Fig. 5.11).

Fig. 5.10 M50 by Anthony Glover

Object Type	Open star cluster
Declination	−8° 20 min
Right Ascension	7 h 3.2 min
Magnitude	5.9

What It Looks Like

M50 can be seen as a misty patch to 70 mm binoculars but starts to "come alive" in a 127 mm Maksutov, when many more stars become resolved. Like many open star clusters, it is irregular in shape but can be imagined as a heart.

Charles Messier's Original Notes

5 April 1772
(RA: 06 h 51 m 50 s, 102d 57' 28"; Dec: −07d 57' 42"; Diameter: –)

Small clusters of stars more or less bright, below the right thigh of the Unicorn [Monoceros], above the star θ of the ear of the Great Dog [Canis Major], & near a star of 7th magnitude. It is by observing the Comet of 1772 that M. Messier observed this cluster. He reported it on the chart of the Comet, which he has drawn. *Memoirs of the Academy for the year 1772.*

How to View

As a personal preference, I like to see M50 with many but not all stars resolved individually. The binocular view is not memorable. It is small enough to take large apertures and the ensuing magnification, as you only need a field of view of about 25 arcminutes.

Photographic Details

This photograph of M50 was taken using an SBIG ST2000XCM cooled to 25° below ambient temperature) with a 80 mm APO Refractor and Astronomik LPR filter. Thirty one-minute subframes were stacked. The view through a 127 mm Maksutov looks very similar (Fig. 5.10).

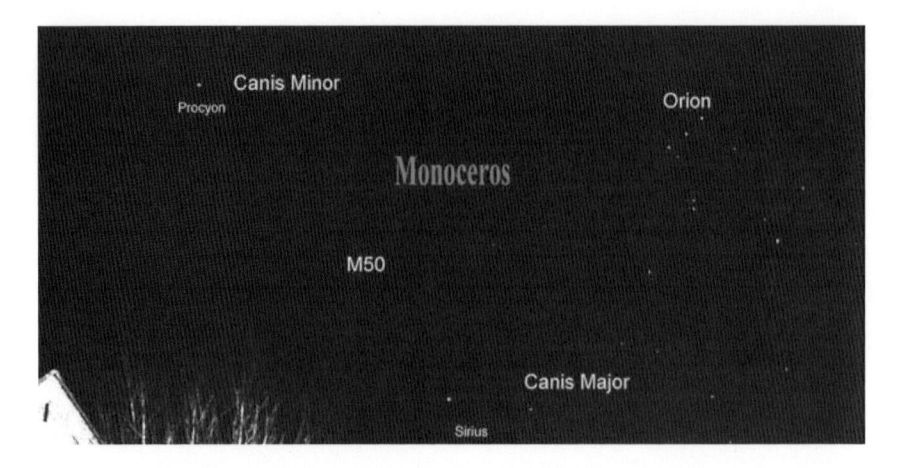

Fig. 5.11 How to find M50

M51

M51 is also known as the Whirlpool Galaxy and is in the constellation of Canes Venatici. It was discovered by Charles Messier in 1773. It also has a faint companion galaxy, NGC5195, and the pair appear to be distorting each other gravitationally. Its apparent dimension is 11×7 arcminutes. Its distance hasn't been well established and estimates vary from 20 million light years to 37 million light years. It can be detected it as a fuzzy patch in 15×70 binoculars. Although it is circumpolar from all of the UK and most parts of North America, it is best observed during spring when it is near the zenith.

Where to Find

M51 is approximately 2° south west of Eta Ursae Majoris. If you draw an imaginary line from the star to Cor Caroli (Alpha Canum Venactorum), it is about a fifth of the way along. As it is quite small, it brightness is more concentrated than the nearby M101, so it will show in a finderscope on a good night. Although you won't get a great view of it, it can be seen through 70 mm binoculars on a good night, so this is a good trick to get you to familiarize yourself with its position (Fig. 5.14).

Fig. 5.12 M51 by Anthony Glover

Object Type	Spiral galaxy
Declination	47° 12 min
Right Ascension	13 h 29.9 min
Magnitude	8.4

What It Looks Like

M51 isn't one of the easier objects in the Messier catalogue and it is barely visible as a misty patch in 70 mm binoculars. Larger apertures certainly liven it up and some spiral structure is visible in a 127 mm Maksutov on a good night and nearby NGC5195 is visible but not so clearly as M51. A light pollution reduction filter somewhat improves the view and brings NGC5195 to some prominence.

Charles Messier's Original Notes

11 January 1774
(RA: 13 h 20 m 23 s, 200d 05' 48"; Dec: +48d 24' 24"; Diameter: –)

[A] very faint nebula, without stars, near the ear of the Greyhound [Canes Venatici], the most northerly, below the star η [Canum Venaticorum] of 2nd magnitude across the tail of Ursa Major: M. Messier discovered this nebula on 13 October 1773, by observing the Comet which appeared this year. It cannot be seen without difficulty with an ordinary telescope of 3½ ft: near it is a star of eighth magnitude. M. Messier reported its position on the chart of the comet observed in 1773 and 1774. *Memoirs of the Academy for the year 1774, plate III.* It is double, each with a bright centre, away from one another by 4' 35". Both atmospheres touch, one is fainter than the other. Reviewed several times.

How to View

Being quite small at 11×7 arcminutes, it can take a little more magnification than many deep sky objects. However, if the conditions are less than perfect, it is best to keep the field of view to at least a degree. It can be seen as a fuzzy patch in modest instruments but the larger the aperture you have available, the better.

Photographic Details

M51 was taken using 5×10 min exposures with an 110 mm/F7 refractor, SBIG St2000×cm, dark subtracted, stacked and processed in *Nebulosity* (Fig. 5.13).

Figure 5.13 was taken using a 10" F5 Newtonian, Starlight Xpress H×916, guided with an 80ED refractor with a modifed Toucam. It consisted of 88 5 min exposure luminance frames and ten 5 min exposures of hydrogen alpha frames (binned 2×2) (Figs. 5.12 and 5.15).

Fig. 5.13 M51 by Mike Deegan

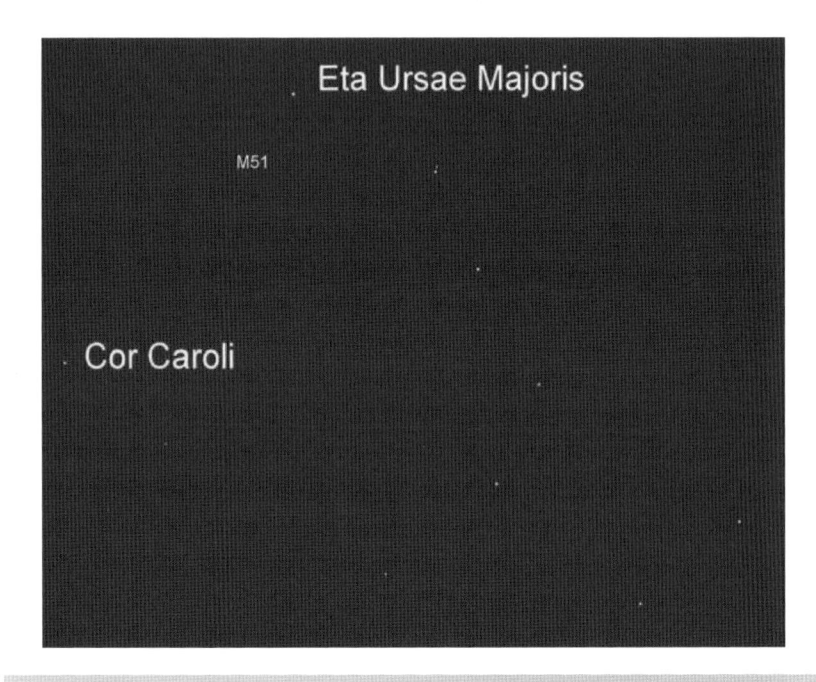

Fig. 5.14 How to find M51

Fig. 5.15 M51 modified

M52

M52 is an open star cluster in Cassiopeia. Its distance isn't well known and estimates vary from 3,000 to 7,000 light years. It is 13 arcminutes across and has about 193 members, the brightest being magnitude 7.77. It was discovered by Charles Messier in 1774. It is circumpolar from the UK and nearly all of the USA but is best seen in fall and winter when it is high in the sky. It is possible to spot in 15×70 binoculars.

Where to Find

M52 is in the northern part of Cassiopeia and can be found by joining an imaginary line between Alpha and Beta Cassiopei and extending it to the north west. It is bright enough to show up in many finderscopes and is easy to find in large binoculars (Fig. 5.16).

Object Type	Open star cluster
Declination	61° 35 min
Right Ascension	23 h 24.2 min
Magnitude	7.3

M52

Fig. 5.16 How to find M52

What It Looks Like

M52 can be seen as a group of stars that stand out against the Milky Way background. It is visible in 70 mm binoculars on a reasonable night and on a good night is obvious in a 50 mm finderscope and, indeed, looks great against the background, which is one of the more sparse regions of the Milky Way. The Maksutov really makes the cluster stand out from the background.

Charles Messier's Original Notes

7 September 1774
(RA: 23 h 14 m 38 s, 348d 39' 27"; Dec: +60d 22' 12"; Diameter: –)

Clusters of very small stars, filtered through with nebulosity, which cannot be seen unless one has an achromatic telescope. In observing the Comet which appeared this year M. Messier saw this cluster, which was near the comet on 7 September 1774 and is below the star d Cassiopeiae: this star was used to determine the cluster of stars and the Comet.

Fig. 5.17 M52 photographed by Anthony Glover

How to View

Aperture fever isn't really necessary with M52. A large aperture will reveal more faint members, but it is a fairly rich star cluster anyway. It was worth the extra effort to find it in a Maksutov, as opposed to the binoculars but only just.

Photographic Details

Figure 5.17 shows M52. It was taken using a 80 mm/F6 refractor and SBIG ST2000×cm camera cooled to −10°. Twenty images, each of 90 s duration were taken and stacked and processed using *Nebulosity* (Stark Software). This figure is a good approximation to how M52 looks through a 127 mm Maksutov.

M53

M53 is a globular star cluster in Coma Berenices. It is a long way from the galactic center both in apparent and real terms and is one of the most distant from us at 58,000 light years. It is 13 arcminutes across. It was discovered by Johann Elert

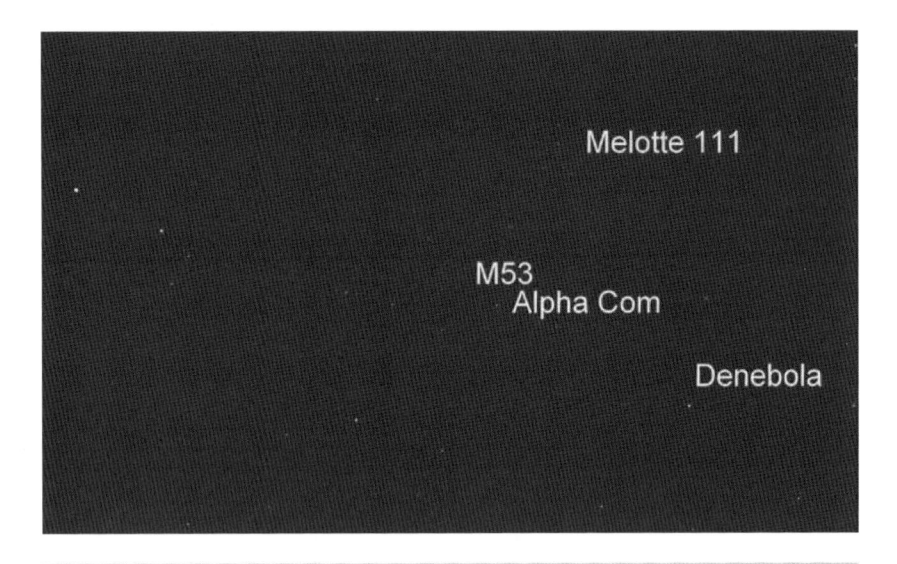

Fig. 5.18 How to find M53

Bode. Through small instruments, it appears identical to an elliptical galaxy, of which there are many in this part of the sky.

Where to Find

M53 lies in a fairly sparse area of the sky but, fortunately, is about a degree north east of Alpha Comae Berenici. Alpha is about halfway between Denebola (Gamma Leonis) and Arcturus. Melotte 111, the well-known open star cluster, is in the same part of the sky (Fig. 5.18).

Object Type	Globular star cluster
Declination	18° 10 min
Right Ascension	13 h 12.9 min
Magnitude	8.2

What It Looks Like

M53 can be seen as a misty patch in a 127 mm Maksutov. Its apparent size appears to be much less than the "official" 13 arcminutes, as only the central concentration of stars is visible. It is one of the less spectacular objects in the Messier catalogue. The structure of a bright nucleus and fainter outer regions is more obvious with a light pollution reduction filter.

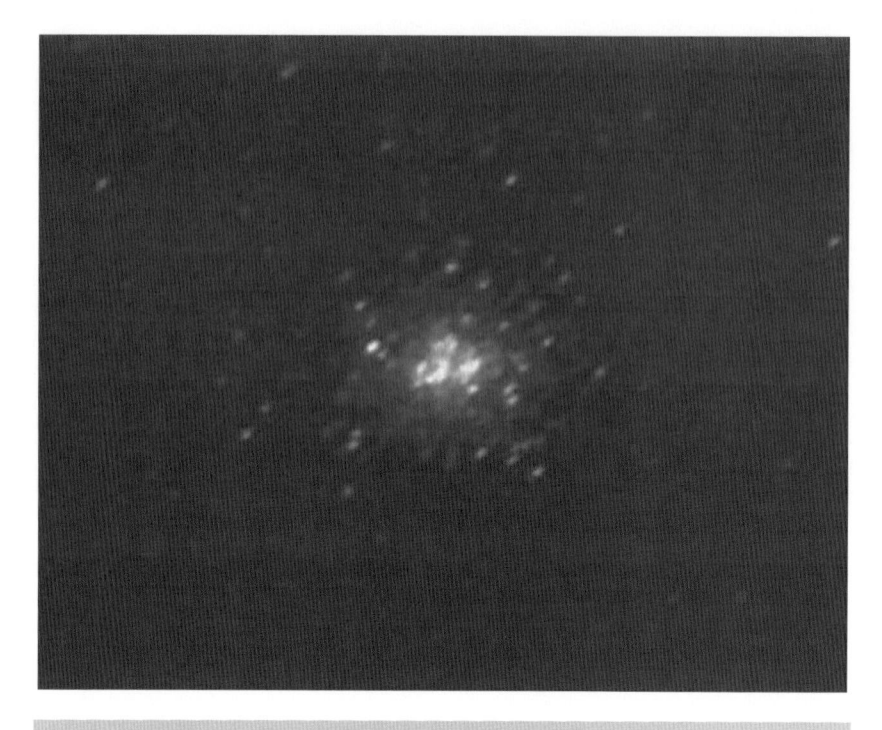

Fig. 5.19 (Courtesy of Kathleen Ollerenshaw, Manchester Astronomical Society)

Charles Messier's Original Notes

26 February 1777
(RA: 13 h 02 m 02 s, 195d 30' 26"; Dec: +19d 22' 44"; Diameter: –)

Nebula without stars discovered below & near the hair of Berenice [Coma Berenices], a short distance from the star 42 of this constellation, according to Flamsteed. This nebula is round & readily-apparent. The Comet of 1779 was compared directly with this nebula, & M. Messier has reported on the chart of the comet, which will be interred in the volume of the Academy of 1779. Reviewed 13 April 1781: it resembles the nebula which is below the Hare [Lepus].

How to View

M53 is unspectacular in small instruments and a large aperture telescope with a field of view of 18 arcminutes or more will show it at its best.

Photographic Details

Figure 5.19

Fig. 5.20 M53 (Courtesy of Kathleen Ollerenshaw, Manchester Astronomical Society)

M54

M54 is a globular star cluster in Sagittarius. It was discovered by Charles Messier in 1778. Although its apparent dimension is 12 arcminutes, most of this brightness is concentrated in the central 2 arcminutes. Messier correctly described it as a "difficult object", as its southerly declination makes it hard to see from the southern USA and impossible from northern Canada and Scotland. It was believed to be about 50,000 light years away, until recent discoveries revealed its true distance of about 87,000 light years. It is even likely that it is not even part of our Milky Way and appears to share a common motion with the Sagittarius Dwarf Elliptical Galaxy.

Due to its great distance, none of its stars is brighter than 15th magnitude and small instruments can only detect the central condensation, at best.

Where to Find

M54 is about 3° east of M70 and 2° north. The nearest bright star is Epsilon Sagittarii (Fig. 5.21).

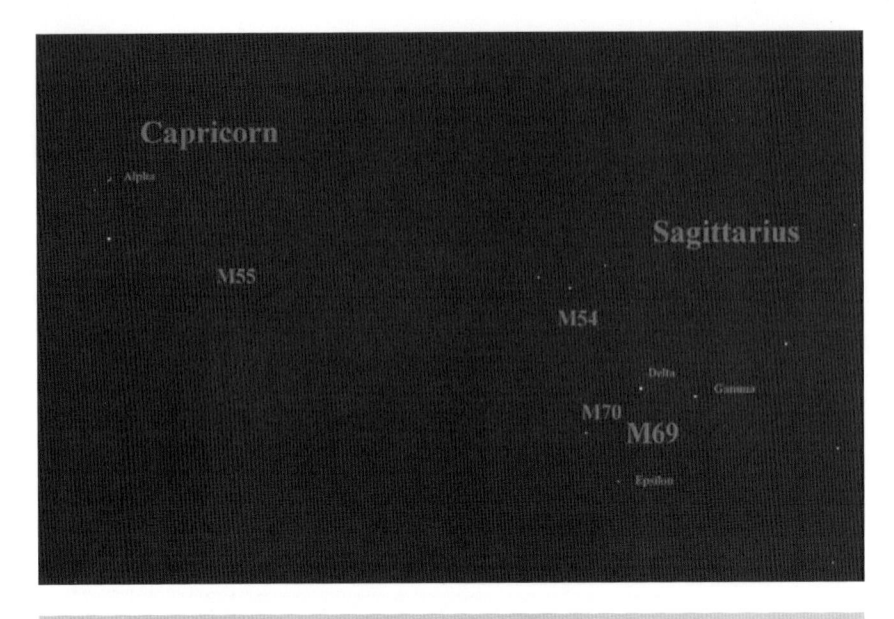

Fig. 5.21 How to find M54

Object Type	Globular star cluster
Declination	−30° 29 min
Right Ascension	18 h 55.1 min
Magnitude	8.0

What It Looks Like

M54 could be a triplet to M69 and M70. From southern skies it probably appears so, although I haven't verified this on visits to the southern hemisphere. The 2° elevation advantage over M70 and M69 actually makes it appear noticeably brighter than them but is still an unspectacular object in binoculars. A 127 mm Maksutov gives a marginally better view but is still unable to resolve individual stars.

Charles Messier's Original Notes

24 July 1778
(RA: 18 h 40 m 52 s, 280d 12′ 55″; Dec: −30d 44′ 01″; Diameter: −)

Fig. 5.22 M54 (NASA/courtesy of nasaimages.org)

Very faint nebula, discovered in Sagittarius, the center is brilliant & contains no star, seen with an achromatic refractor of three feet and a half. Its position was determined from ζ Sagittarii, of 3rd magnitude.

How to View

From a favourable location, a 250 mm aperture telescope is needed to resolve any individual stars. From mid northern latitudes, extinction will mean that twice this aperture is needed. With smaller apertures, it is unremarkable.

Photographic Details

Figures 5.22 and 5.23

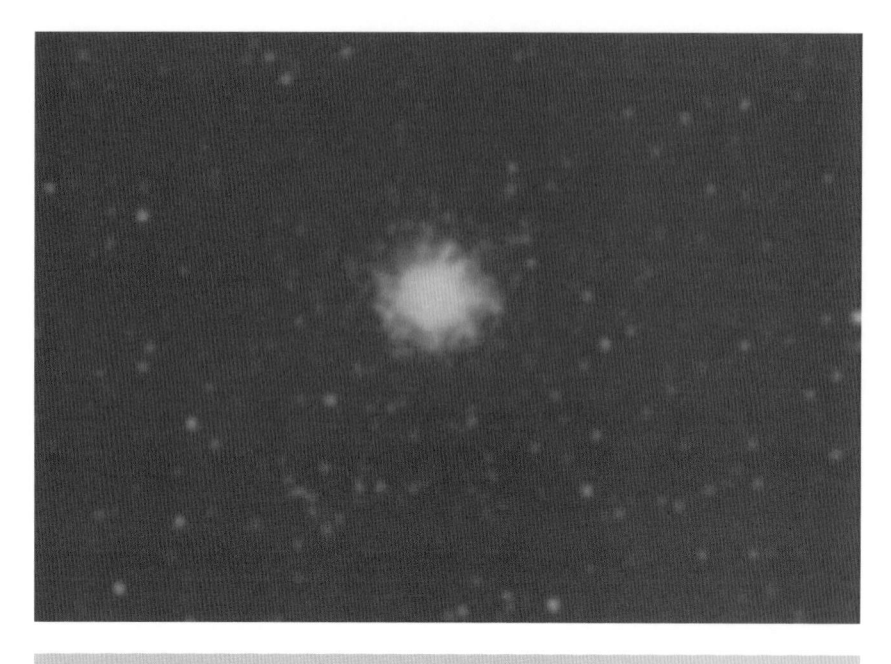

Fig. 5.23 M54 modified (NASA/courtesy of nasaimages.org)

M55

M55 is a globular star cluster in eastern Sagittarius, near the border with Capricorn. It was discovered by Abbe Lacaille from South Africa, where it is a much easier object than from the United States or Europe. It is described as being a loose cluster and is large at 19 arcminutes diameter.

Where to Find

M55 is not near any obvious landmark (or should that be skymark?). It is best found by doing a tour starting at M22 and going south to M70 (perhaps diverting via M69) then north and east to M54. It is then about 8° east and a degree south of M55 (Fig. 5.24).

Object Type	Nebula
Declination	−30° 58 min
Right Ascension	19 h 39.9 min
Magnitude	7.4

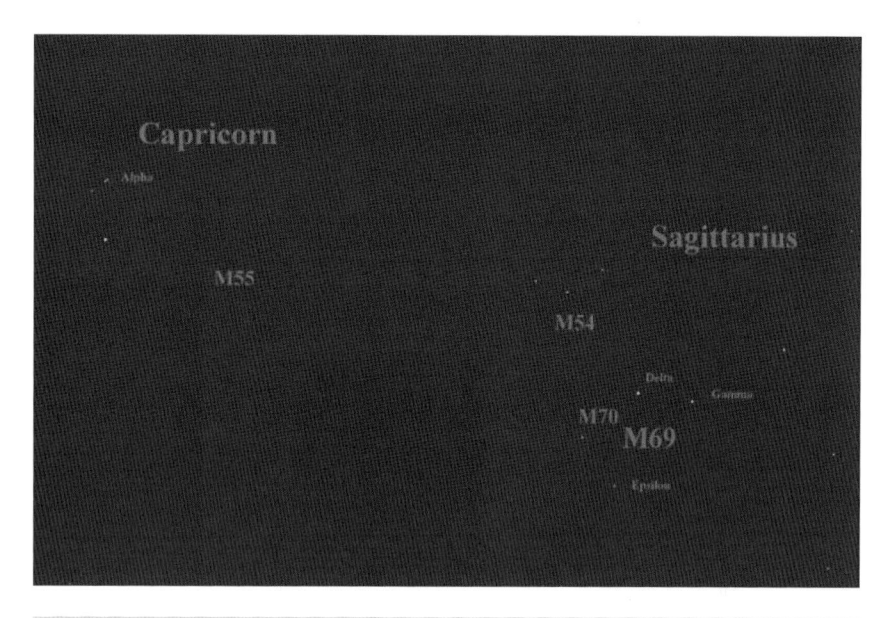

Fig. 5.24 How to find M55

What It Looks Like

Through 70 mm binoculars, M55 looks nothing like its "official" description of being somewhat "grainy" or loose but just appears as a fainter, slightly smaller version of M22. Without modern knowledge, there would be no way of distinguishing it from an elliptical galaxy. A 127 mm Maksutov armed with a light pollution reduction filter brings out the grainy, globular structure a bit better.

Charles Messier's Original Notes

24 July 1778
(RA: 19 h 26 m 02 s, 291d 30' 25"; Dec: −31d 26' 27"; Diameter: −)

Nebula which is a whitish spot, about 6 min in extent, its light is even and doesn't appear to contain any star. Its position was determined from ζ Sagittarii, using an intermediate star of 7th magnitude This nebula was discovered by M. l'Abbé de la Caille. *Memoirs of the Academy for the year 1755, page194*. Messier has sought [it] in vain on 29 July 1764, as he relates in his memoir.

How to View

Under average viewing conditions, M55 is a borderline object for my set-up but can benefit from a larger aperture instrument.

Photographic Details

Figures 5.25 and 5.26

Fig. 5.25 M55 (NASA/courtesy of nasaimages.org)

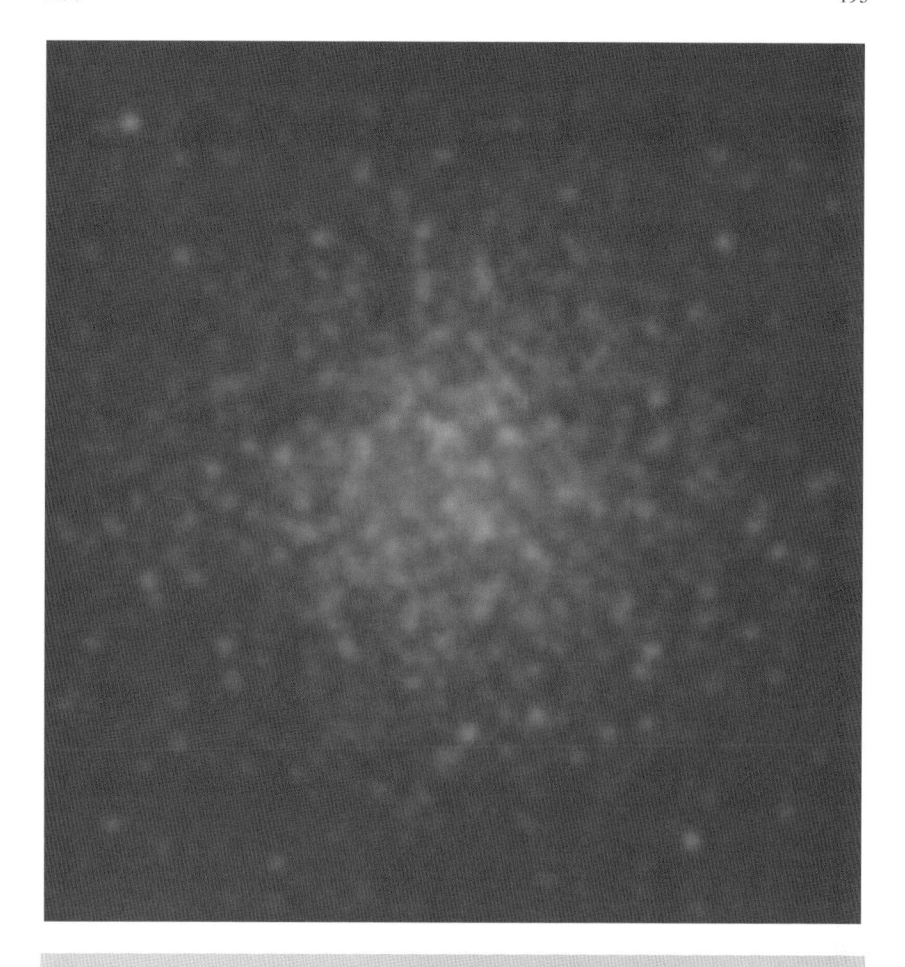

Fig. 5.26 M55 modified (NASA/courtesy of nasaimages.org)

M56

M56 is a globular star cluster in Lyra. Although its recorded dimension is nearly 9 arcminutes, only the central 3 arcminutes is usually seen. It is nearly 33,000 light years away and its brightest stars are of 13th magnitude, so not visible in small and medium aperture instruments. It was discovered by Charles Messier in 1779. It is comparable in brightness to the Ring Nebula (M57).

Fig. 5.27 M56 by Anthony Glover

Where to Find

Although M56 is within the borders of Lyra, the nearest bright star is Albireo (Beta Cygni), the well known double star. It is about 3° north west of it (Fig. 5.28).

Object Type	Global star cluster
Declination	30° 11 min
Right Ascension	19 h 16.6 min
Magnitude	8.3

What It Looks Like

M56, being an 8th magnitude globular star cluster, does not look much different to an elliptical galaxy. Some of its outer regions can be resolved in a 127 mm Maksutov. It is barely visible in 70 mm binoculars.

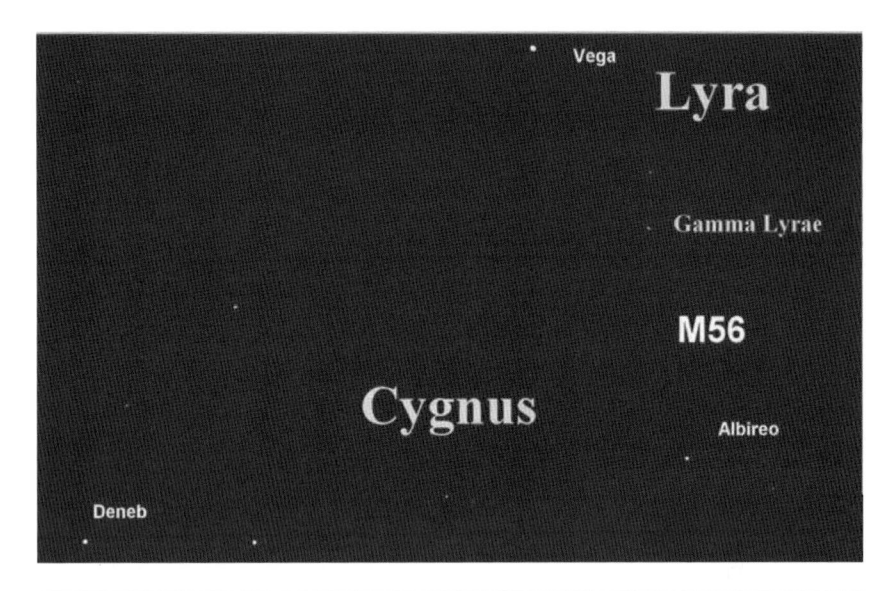

Fig. 5.28 How to find M56

Charles Messier's Original Notes

23 January 1779
(RA: 19 h 08 m 00s, 287d 00' 01"; Dec: +29d 48' 14"; Diameter: –)

Nebula without stars, with little light; M. Messier [also] discovered on the same day the Comet of 1779 on January 19. On 23 [January] its position was determined by comparing it to the star two of the Swan [Cygni], according to Flamsteed: it is near the Milky Way; it is near to a star of 10th magnitude. M. Messier has reported it on the chart the Comet of 1779.

How to View

M56 is best viewed using apertures of at least 200 mm.

Photographic Details

M56 was taken using a Skywatcher ED100, semi-APO, SXV H9, with an infrared blocking filter. It consisted of 16 1 min exposures (Figs. 5.27 and 5.29).

Fig. 5.29 M56 modified

M57

M57 is also known as the Ring Nebula. It was discovered by Antoine Darquier de Pellepoix in 1779, only a few days before Charles Messier. Although it is fainter than the Dumbbell (M27), it is better known as it is easier to find. Its apparent size is 1.4×1 arcminutes, so it is one of the few deep sky objects that responds to high magnification. Its exact nature is unknown, as scientists don't know whether it is a torus (or doughnut shape) or a cylinder. Estimates of its distance vary but a better idea was discovered recently by estimating the distance to the central star of 2,300 light years. This central star is a 15th magnitude white dwarf, which will be the ultimate fate of our Sun.

It is not one of the easiest deep sky objects and it is borderline with my 15×70 binoculars on all but exceptional nights and most nights it is beyond them. It is an interesting target for my 127 mm Maksutov-Cassegrain, although one night I just couldn't find it, despite "bagging" it the first night after the telescope arrived. Its high northern declination helps its appearance from the UK, USA and Canada but it is a challenge from 35° south of the equator, where it is a winter object.

Where to Find

(Figure 5.31)

Fig. 5.30 M57 by Anthony Glover

How to Find

M57 is in the small but distinct constellation of Lyra between Gamma and Beta Lyrae but somewhat closer to Gamma.

Object Type	Planetary nebula
Declination	33° 02 min
Right Ascension	18 h 53.6 min
Magnitude	8.8

What It Looks Like

M57 looks like a fuzzy patch through binoculars or a small telescope on a clear night. A 127 mm Maksutov with moderate to high magnification (about 200×) shows the ring structure but about 250–300 mm of aperture is needed to see the central star. It really looks like it does in the photographs (and M57 is a very photographed object), except that visual observation doesn't pick up the colors.

Charles Messier's Original Notes

31 January 1779
(RA: 18 h 45 m 21 s, 281d 20' 08"; Dec: +32d 46' 03"; Diameter: –)

Clusters of light placed between β and γ of the Lyre [Lyrae], discovered when observing the Comet of 1779, which passed very close by: it seemed that this cluster of light, which is round, was composed of very small stars: with best telescopes you cannot see them, it remains only a suspicion that they are there. M. Messier has reported this cluster of light on the chart of the Comet of 1779. M. Darquier, in Toulouse, discovered the nebula, whilst observing the same comet, and he reports: "Nebula between β and γ Lyrae; which is very dull, but perfectly defined; it is as big as Jupiter and resembles a planet which is extinguishing."

How to View

The great thing about M57 is that it can be enjoyed in any sort of instrument from 100 mm aperture up. When well-placed, a light pollution reduction filter doesn't help much but it does when extinction and murk are around when it is viewed "out of season" at low elevations. It is best placed in July and August. Although large apertures help, especially if one wishes to see the central star, they are not totally necessary to obtain the "wow" factor.

Photographic details

Figures 5.30 and 5.32.

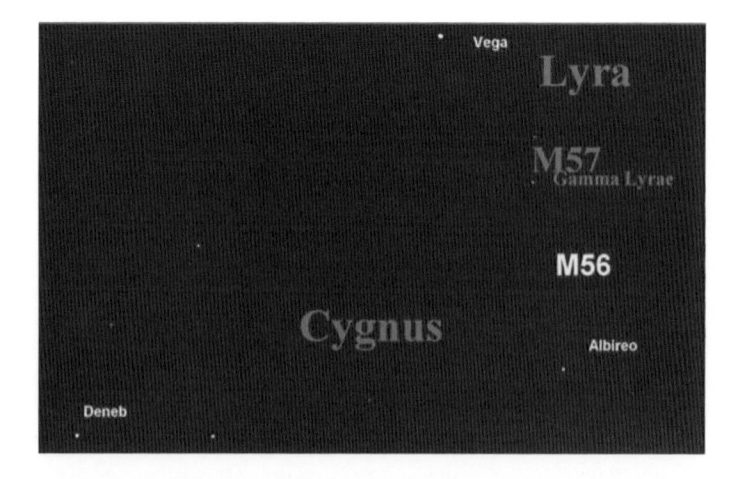

Fig. 5.31 How to find M57

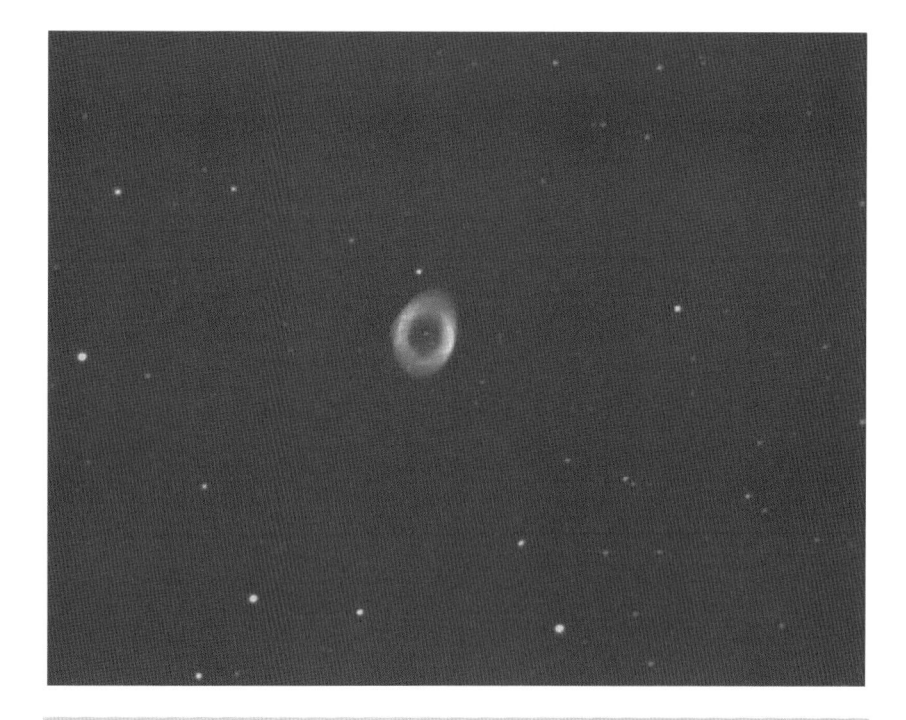

Fig. 5.32 M57 modified

M58

M58 is one of the members of the Virgo Supercluster and is about 60 million light years away. It was discovered by Charles Messier in 1779 and is 4.5×5.5 arcminutes in apparent size. Its spiral structure is not apparent in small instruments, as only its bright core can be seen, suggesting that it is an elliptical galaxy, although it is actually a barred spiral galaxy.

Where to Find

M58 is in the densest part of the Virgo Supercluster. At magnitude 9.7, it is rather faint and can be missed on days when the nearby brighter M60 and M87 can be seen. It is about a degree and a half west of M60 and slightly to the north. It is best to use Epsilon Virginis and M60 as guides and if M60 is invisible, then M58 will be as well.

Object Type	Barred spiral galaxy
Declination	11° 49 min
Right Ascension	12 h 37.7 min
Magnitude	9.7

What It Looks Like

M58 is one of the fainter Messier objects and is a borderline object for a 127 mm Maksutov. Under suburban conditions, it appears as a misty patch where a galaxy should be. Although it is a barred spiral galaxy, it is impossible to determine this with a modest instrument, so is best regarded as "one to cross off the list".

Charles Messier's Original Notes

15 April 1779
(RA: 12 h 26 m 30s, 186d 37' 23"; Dec: +13d 02' 42"; Diameter: –)

Nebula, very faint, discovered in Virgo, almost along the parallel of ε [Virginis], 3rd magnitude. No light can be used to illuminate the micrometer as it [the nebula] disappears. M. Messier has reported it on the chart of the Comet of 1779, which will be in the volume of the Academy for that same year.

How to View

Quite simply, M58 is not a suitable target for modest instruments and its structure is more readily revealed in telescopes with an aperture of 250 mm or more.

Photographic Details

Figures 5.34

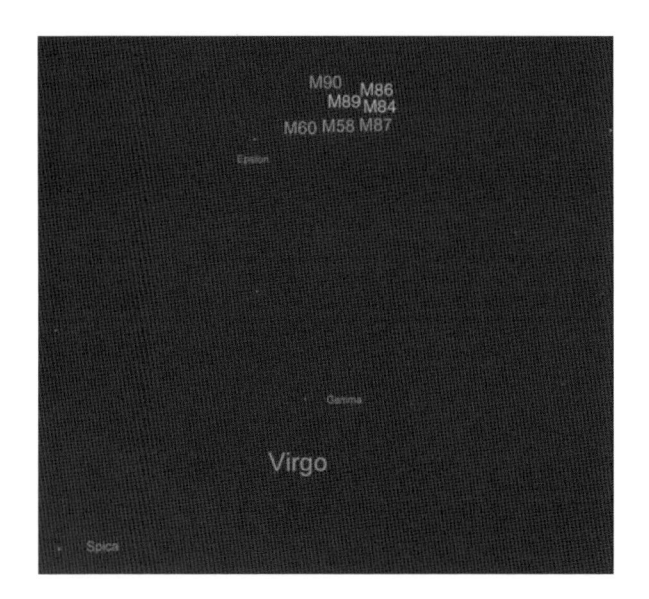

Fig. 5.33 How to find M58

Fig. 5.34 M58 modified (NASA/courtesy of nasaimages.org)

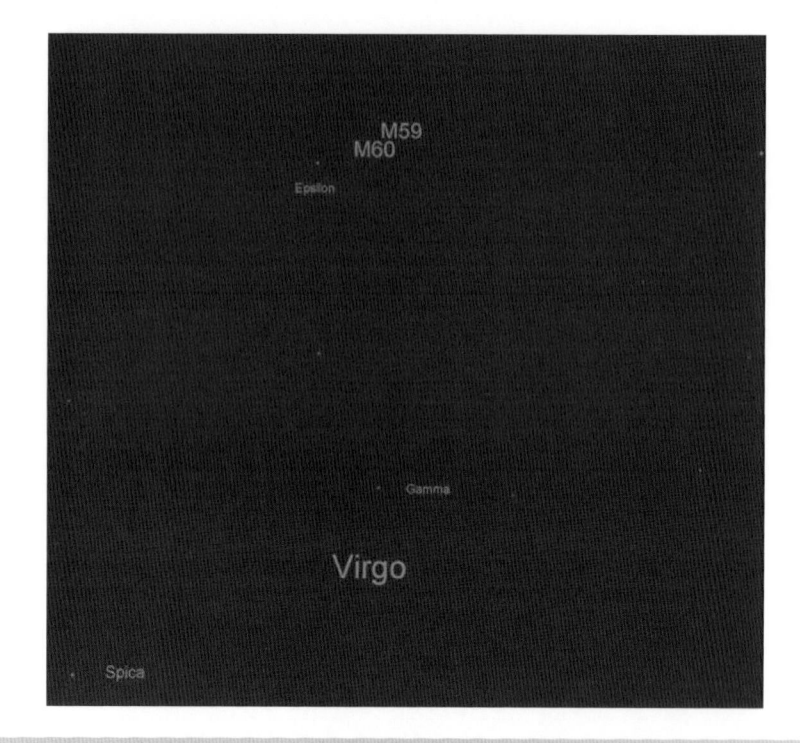

Fig. 5.35 How to find M59

M59

M59 is a large elliptical galaxy in the Virgo Supercluster, suggesting a distance of about 60 million light years. It has dimensions of 5×3.5 arcminutes, suggesting a major axis of 90,000 light years and a minor axis of 55,000 light years. M59 was discovered by Johann Gottfried Koehler on April 11, 1779.

Where to Find

M59 is one of the "bowl" galaxies of Virgo. It is in a fairly sparse area of sky and about 3° west and a degree north of Epsilon Virginis. M60 is a close neighbor and is somewhat brighter. Without accurate finding equipment or techniques, it is likely that you will find M60 first, which is a borderline target for finderscopes on a perfect night under dark sky conditions. From a suburban location, finding M59 and M60 is rather "fiddly" (Fig. 5.35).

Object Type	Elliptical galaxy
Declination	11° 39 min
Right Ascension	12 h 42.0 min
Magnitude	9.6

What It Looks Like

M59 is on the border of resolution with a 127 mm Maksutov under good conditions, so just appears as a faint fuzzy object in the vicinity of M60. There is little to suggest its true nature, without darker skies and/or larger instruments. However, it is more prominent in a light pollution reduction filter.

Charles Messier's Original Notes

15 April 1779
(RA: 12 h 30 m 47 s, 187d 41′ 38″; Dec: +12d 52′ 36″; Diameter: –)

Nebula in Virgo & in the vicinity of the preceding one, on the [same] parallel of ε [Virginis], which served as its determination [of position]: it has the same light as the one above, also as faint. M. Messier has reported the chart of the Comet of 1779.

How to View

Although a light pollution reduction filter helps, M59 is not a suitable target for modest instruments and requires it to be placed near the meridian for viewing. More detail is visible in larger aperture instruments.

Photographic Details

Figures 5.36 and 5.37

Fig. 5.36 M59 (NASA/courtesy of nasaimages.org)

Fig. 5.37 M59 modified (NASA/courtesy of nasaimages.org)

M60

Like its neighbor, M59, M60 was discovered 1779 by Johann Gottfried Koehler. It is a truly large, massive elliptical galaxy in the Virgo Supercluster (making it about 60 million light years away), having dimensions of 7×6 arcminutes and a core of 4×3 arcminutes, visible to small instruments. It has an actual size of about 120,000 light years across, making it larger than our own Milky Way and it has a massive object of two billion solar masses in its core, probably a black hole.

Where to Find

Please refer to Fig. 5.35 as M60 is very close to M59.

Object Type	Elliptical galaxy
Declination	11° 33 min
Right Ascension	12 h 43.7 min
Magnitude	8.8

What It Looks Like

Whilst M60 is not a spectacular object in modest instruments, nevertheless it is large and bright enough for a 127 mm Maksutov to show it to be an elliptical object of indeterminate nature, suggesting an elliptical galaxy (its true nature) or a globular star cluster. However, it is more prominent in a light pollution reduction filter and its nature as an elliptical galaxy is more obvious than without.

Charles Messier's Original Notes

15 April 1779
(RA: 12 h 32 m 28 s, 188d 06′ 53″; Dec: +12d 46′ 02″; Diameter: –)

Nebula in Virgo, a little more apparent than the previous two, also on the [same] parallel of ε [Virginis], which was used for its determination [of position]. M. Messier has reported it on the chart of the Comet of 1779. He discovered these three nebulae whilst observing this comet which passed by very near them. The last passed so near on the 13 and 14 of April, that one and another were in the same field [of view] of the telescope, and he could not see it; it was only on 15 [of April], looking for the comet, that he saw this nebula. These three nebulae don't contain any star[s].

Fig. 5.38 M60 (NASA/courtesy of nasaimages.org)

How to View

Although M60 is discernable as an elliptical galaxy in a 127 mm Maksutov and light pollution reduction filter, it is best viewed with larger aperture instruments.

Photographic Details

Figures 5.38 and 5.39

Fig. 5.39 M60 modified (NASA/courtesy of nasaimages.org)

M61

M61 is another member of the Virgo Supercluster at a distance of about 60 million light years. It was discovered by Barnabus Oriani in 1779. Charles Messier actually saw it on the same day but mistook it for a comet that was known to be in the vicinity, rather like me seeing M92 on a clear night and getting confused. Messier, however, did not have the type of star atlas that is available today. Its apparent size of 6×5.5 arcminutes corresponds to a real size of 100,000 light years, about the same as our Milky Way.

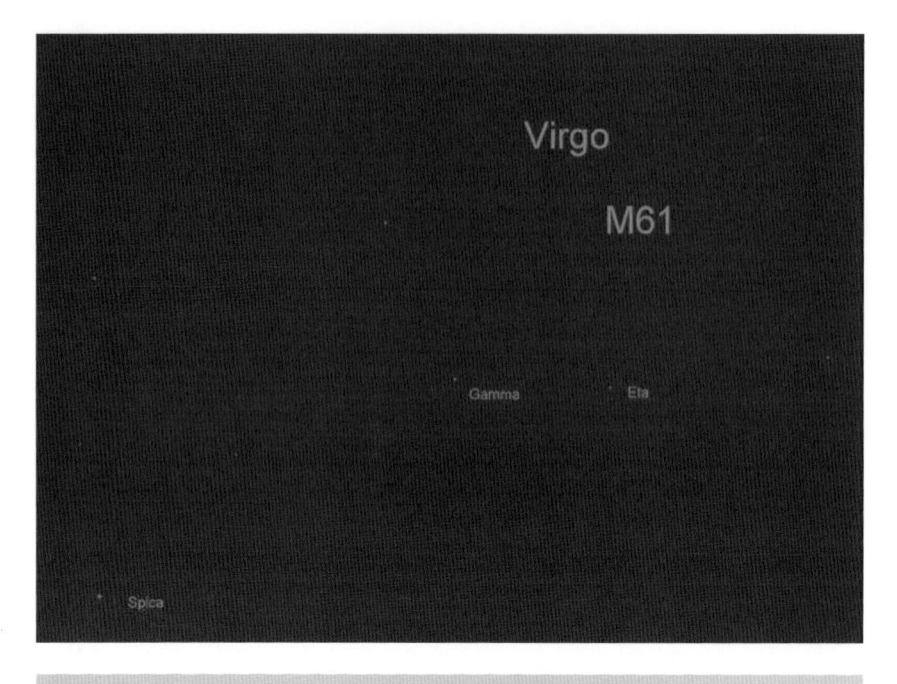

Fig. 5.40 How to find M61

Where to Find

M61 is one of the hardest Messier objects to find. Although some are fainter, it never gets particularly high from northern Europe and there are no easy guide objects. It is about 5° north of Eta Virginis and lies between the fainter stars 16 and 17 Virginis, although neither are visible in Fig. 5.40.

Object Type	Spiral galaxy
Declination	4° 28 min
Right Ascension	12 h 21.9 min
Magnitude	9.7

What It Looks Like

M61 appears as a faint patch of light in a 127 mm Maksutov with no obvious shape. It is near the limit of resolution and, being one of the southernmost members of the main Virgo Supercluster, suffers from extinction.

Fig. 5.41 M61 (NASA/courtesy of nasaimages.org)

Charles Messier's Original

11 May 1779
(RA: 12 h 10 m 44 s, 182d 41′ 05″; Dec: +05d 42′ 05″; Diameter: –)

Nebula, faint and very difficult to see. M. Messier mistook this nebula for the Comet of 1779, on 5th, 6th & 11th May; on the 11th he recognized that this was not the comet, but a nebula which was on its path & at the same point in the sky.

How to View

M61 is not a suitable target for small instruments and is more one to "cross off the list" than expect to see anything exciting. It is better to use a large aperture instrument.

Photographic Details

Figures 5.41 and 5.42

Fig. 5.42 M61 modified (NASA/courtesy of nasaimages.org)

M62

M62 is a globular star cluster in Ophiuchus. It was discovered by Charles Messier in 1771. It is 15 arcminutes across and 22,500 light years away. Its southerly declination makes it a difficult object from even southern England and Canada.

Where to Find

M62 lies in a rather sparse part of sky, with no obvious bright stars or other landmarks. The best way to find it is to find M19 then move 4° south (Fig. 5.44).

Object Type	Globular star cluster
Declination	−30° 07 min
Right Ascension	17 h 01.2 min
Magnitude	6.5

Fig. 5.43 M62 (NASA/courtesy of nasaimages.org)

What It Looks Like

M62 should be visible to 70 mm binoculars on a clear night but it wasn't from southern England. With that in mind, one would anticipate that it would appear as a mere faint fuzzy in a larger instrument. Fortunately, it was a nice surprise, showing a definite central condensation of stars, suggesting its nature was a globular star cluster or distant galaxy.

Charles Messier's Original Notes

4 June 1779
(RA: 16 h 47 m 14 s, 251d 48' 24"; Dec: −29d 45' 30"; Diameter: −)

Very beautiful nebula, discovered in the Scorpion [Scorpio], it looks like a small comet, the centre is brilliant & surrounded by a faint glow. Its position determined by comparing it to the star τ. M. Messier has already seen this nebula on 7th June 1771, without having determined the position it is close to. Reviewed on 22 March 1781.

How to View

Evidence would suggest that an instrument of at least 70 mm of aperture should be needed to see M62 from mid-northern latitudes. Whilst large apertures are not essential, as for most globular star clusters, they help resolve some of the stars. A light pollution reduction filter is a great help, especially, as it never rises more than 10° anywhere in the UK.

Photographic Details

Figures 5.43 and 5.45

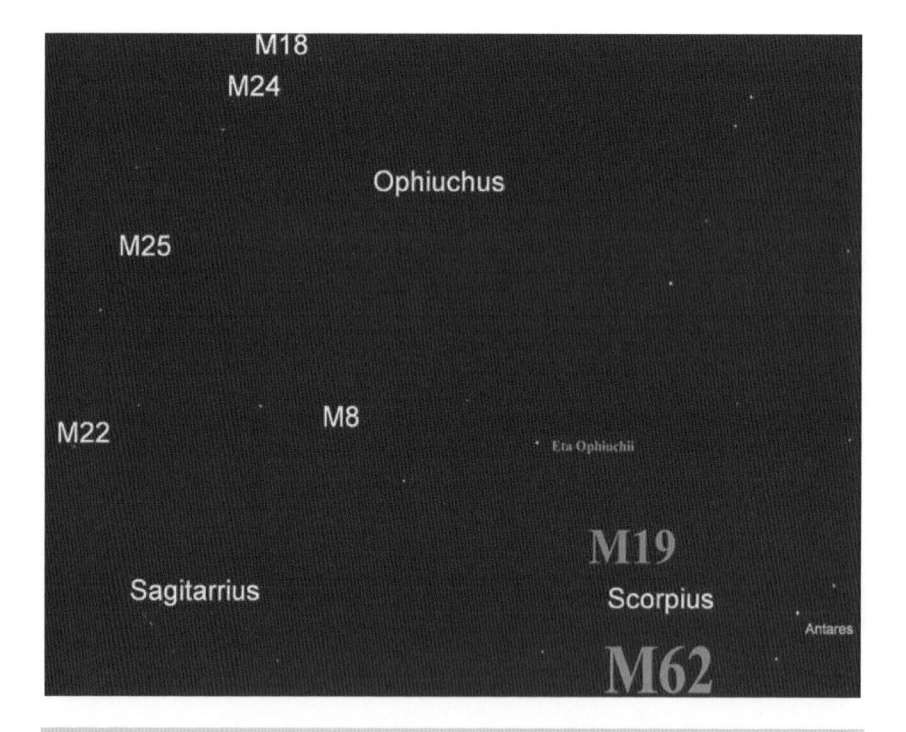

Fig. 5.44 How to find M62

Fig. 5.45 M62 modified (NASA/courtesy of nasaimages.org)

M63

M63 was discovered 1779 by Pierre Méchain. It is a spiral galaxy in Canes Venatici, which has a dimension of 10×6 arcminutes and distance of 37 million light years. It is also known as the Sunflower Galaxy. It appears to be physically associated with the nearby Whirlpool Galaxy (M51). It is circumpolar from the UK and Canada.

Where to Find

M63 lies roughly on a line between Alkaid (Eta Ursae Majoris) and Cor Caroli (Alpha Canum Venactorum). It is about a third of the way between the two stars. The difficulty is that it is too faint to see in finderscopes (Fig. 5.47).

Fig. 5.46 M63 by Anthony Glover

Object Type	Spiral galaxy
Declination	42° 02 min
Right Ascension	13 h 15.8 min
Magnitude	8.6

What It Looks Like

M63 appears as a fuzzy patch to 70 mm binoculars but looks nice with a 127 mm Maksutov armed with a light pollution reduction filter. It appeared somewhat like the Pinwheel (M33) but with a flatter shape. Surprisingly enough, the nucleus was not as obvious as the spiral arms.

Charles Messier's Original Notes

14 June 1779
(RA: 13 h 04 m 22 s, 196d 05' 30"; Dec: +43d 12' 37"; Diameter: –)

Nebula discovered by M. Méchain in the Greyhound [Canes Venatici]. M. Messier has looked for it; it is faint, it has almost the same light as the nebula reported as

No. 59 [M59]: it contains no star, & the slightest light to illuminate the micrometer makes it disappear: there is close-by a star of 8th magnitude, that precedes the nebula over time. M. Messier has reported its position on the chart of the Comet of 1779.

How to View

Whilst M63 is not one of the brighter Messier galaxies, here the key is to view it when it is near the zenith. I would suggest that my Maksutov is about the minimum aperture required to make out the spiral arms and the light pollution reduction filter was a definite help.

Photographic Details

Figure 5.46 shows M63. It was taken using a 80 mm/F6 refractor and SBIG ST2000×cm camera cooled to −25°. Fifteen images, each of 5 min duration were taken and stacked and processed using *Nebulosity* (Stark Software). This figure is a good approximation to how M63 looks through a 127 mm Maksutov.

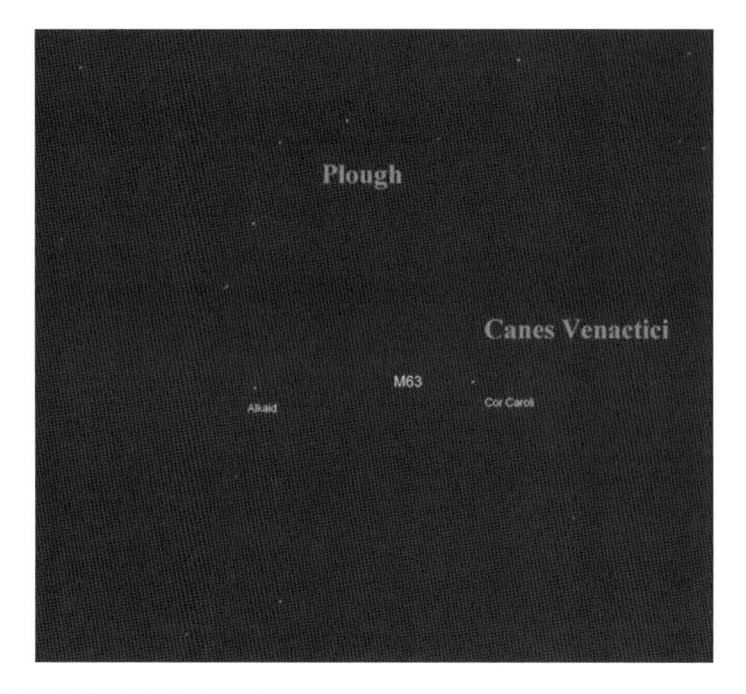

Fig. 5.47 How to find M63

M64

M64 is a spiral galaxy in Coma Berenices, also known as the Black Eye Galaxy or Sleeping Beauty Galaxy. It was discovered in 1779 by Edward Pigott. Its distinguishing feature is a dark dust cloud that is in the outer reaches of the galaxy and obscures stars behind it. It has an apparent size of 9.3×5.4 arcminutes and is about 19 million light years away, according to the most recent estimates. Its motion is probably affected by the Virgo Supercluster.

Its magnitude and northerly declination make it an interesting target for medium aperture instruments during spring. It is believed to be physically associated with some nearby small irregular galaxies and possibly with those in Canes Venatici.

Where to Find

M64 is in the same approximate neighbourhood as M53 but is a lot fainter. It is quite a good idea to go for both objects in the same session. Thirty Five Comae Berenici is between the open star cluster Melotte 111 and Alpha Comae Berenici. M64 is just under a degree from 35 on a bearing of approximately east north east (Fig. 5.49).

Fig. 5.48 M64 by Mike Deegan

M64

Object Type	Spiral galaxy
Declination	21° 41 min
Right Ascension	12 h 56.7 min
Magnitude	8.5

What It Looks Like

M64, at first glance, just looks like a misty patch through a 127 mm Maksutov but closer inspection reveals its spiral structure and it is reminiscent of the Pinwheel (M33) as seen through binoculars. Although it is not as spectacular as some of the brighter Messier objects, nevertheless, it actually does look like a spiral galaxy, although the dark dust lane was not visible and would need a larger instrument and/or clearer sky to resolve.

Charles Messier's Original Notes

1 March 1780
(RA: 12 h 45 m 51 s, 191d 27′ 38″; Dec: +22d 52′ 31″; Diameter: –)

Nebula discovered in the hair of Berenice [Coma Berenices], which is less than half as apparent as that which is below the hair. M. Messier has reported its position on the chart of the Comet of 1779. Reviewed on 17 March 1781.

How to View

Whilst M64 is not one of the brighter Messier galaxies, here the key is to view it when it is near the zenith. I would suggest that my Maksutov is about the minimum aperture required to make out the spiral arms and the light pollution reduction filter was a definite help.

Photographic Details

Figures 5.48 and 5.50

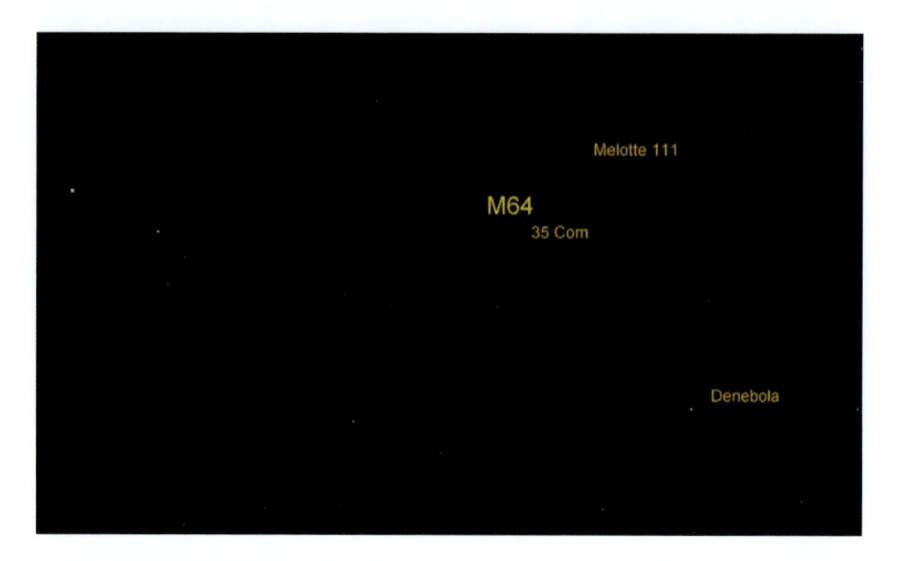

Fig. 5.49 How to find M64

Fig. 5.50 M64 Modified

M65

M65 is a spiral galaxy in Leo, discovered by Charles Messier in 1780. It has dimensions of 8×1.5 arcminutes, as it is not far from being edge on. It forms a close group with M66 and NGC3623, known as the Leo Triplet. Although somewhat fainter than many members of the Virgo Supercluster, it is easier to spot from the northern hemisphere, due to its declination. It is 35 million light years away. It consists of old stars but also has many active star-forming regions.

Where to Find

First find Theta Leonis which is the lowest star in Leo's hind quarters. About 3° due south is 73 Leonis. M65 and M66 are about a degree east of it and can be seen in the same wide field of view. Unfortunately, both are too faint to see in a finderscope (Fig. 5.51).

Object Type	Spiral galaxy
Declination	13° 05 min
Right Ascension	11 h 18.9 min
Magnitude	9.3

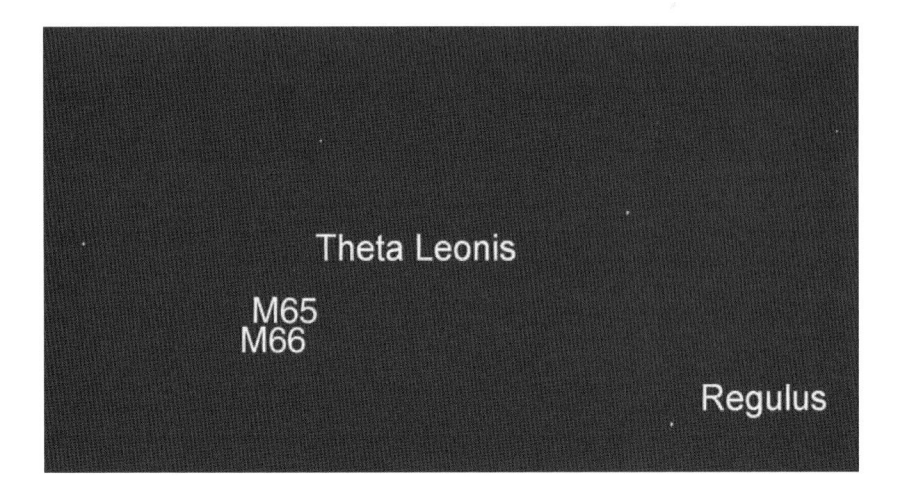

Fig. 5.51 How to find M65

What It Looks Like

M65 appears like a fuzzy patch to big binoculars but you can't tell what it is. I've been fortunate enough to see M65 and M66 through an 8″ Schmidt-Cassegrain and they were superb. My 127 mm Maksutov-Cassegrain shows both galaxies as spiral ones but rather like a binocular view of the Andromeda Galaxy (M31) on a poor night.

Charles Messier's Original Notes

1 March 1780
(RA: 11 h 07 m 24 s, 166d 50′ 54″; Dec: +14d 16′ 08″; Diameter: –)

Nebula discovered in Leo, it is very faint & contains no star.

How to View

Although I've picked it up in binoculars, it really needs a telescope and, the larger the aperture, the better.

Photographic Details

Figure 5.52, showing the Leo Triplet (M65, M66 and NGC3628) was taken using ten luminance exposures of 10 min using the ED80 refractor combined with ten 5 min exposures for red and seven 5 min exposures for each of red and green for the starfield background.

NGC3628 was taken using a 10″ F5 Newtonian, Starlight Xpress H×916, guided with a 80ED refractor with a modifed Toucam. It consisted of 37 5 min exposure luminance frames, 10 150 s RGB exposures binned 2×2 and ten 5 min exposures of hydrogen alpha.

M66 was taken using a 10″ F5 Newtonian, Starlight Xpress H×916, guided with a 80ED refractor with a modifed Toucam. It consisted of 24 5 min exposure luminance frames, 10 90 s RGB exposures binned 2×2 and 12 5 min exposures of hydrogen alpha binned 2×2.

M65 was taken using a 10″ F5 Newtonian, Starlight Xpress H×916, guided with a 80ED refractor with a modifed Toucam. It consisted of 24 5 min exposure luminance frames, nine 150 s red exposures binned 2×2, fifteen 150 s each of blue and green exposures binned 2×2 and ten 5 min exposures of hydrogen alpha binned 2×2 and a further set of hydrogen alpha frames at half the exposure times (Figs. 5.52 and 5.53).

Fig. 5.52 Leo Triplet by Mike Deegan

Fig. 5.53 M65 modified

M66

M66 is a spiral galaxy in Leo, discovered by Charles Messier in 1780. It has dimensions of 8 × 2.5 arcminutes. It forms a close group with M65 and NGC3623, known as the Leo Triplet. Although somewhat fainter than many members of the Virgo Supercluster, it is easier to spot from the northern hemisphere, due to its declination. It is 35 million light years away. It is noticeably larger than its neighbour M65 and, although brighter, has a similar surface brightness, due to its larger area. All members of this group appear to be warped because of gravitational interaction.

Where to Find

First find Theta Leonis which is the lowest star in Leo's hind quarters. About 3° due south is 73 Leonis. M65 and M66 are about a degree east of it and can be seen in the same wide field of view. Unfortunately, it is too faint to see in a finderscope.

Object Type	Spiral galaxy
Declination	12° 59 min
Right Ascension	11 h 20.2 min
Magnitude	8.9

What It Looks Like

M66 appears like a fuzzy patch to big binoculars but you can't tell what it is. I've been fortunate enough to see M65 and M66 through an 8″ Schmidt-Cassegrain and they were superb. My 127 mm Maksutov-Cassegrain shows both galaxies as spiral ones but rather like a binocular view of the Andromeda Galaxy (M31) on a poor night.

Charles Messier's Original Notes

1 March 1780
(RA: 11 h 08 m 47 s, 167d 11′ 39″; Dec: +14d 12′ 21″; Diameter: –)

M66

Nebula discovered in Leo; with a very faint light & very near the previous [nebula]: they appear together in the same field of the telescope. The Comet observed in 1773 & 1774 has passed between these two nebulae on 1 to 2 November 1773. M. Messier did not see them then, no doubt, because of the light of the comet.

How to View

Although I've picked it up in binoculars, it really needs a telescope and, the larger the aperture, the better.

Photographic Details

Please refer to Fig. 5.52 showing the Leo Triplet of galaxies for an image of M66. Figure 5.54 shows how M66 looks through my 127 mm Maksutov.

Fig. 5.54 M66 modified

M67

M67 is an open star cluster in Cancer and is usually overlooked, due to its closeness to the Beehive (M44). It is a large cluster, being 30 arcminutes across and is quite far away, at 2,700 light years. It was discovered some time before 1779 by Johann Gottfried Koehler. Of all the open star clusters in the Messier Catalogue, it is certainly the oldest with the youngest estimate being 3,200 million years, quite remarkable for an open star cluster. It has about 500 members and some of them are of magnitude 10 and are hot blue stars probably formed by star mergers and are therefore known as "blue stragglers", which are more commonly found in globular star clusters.

Where to Find

M67 is harder to find than M44. Not only does it share the same faint host constellation but it is also noticeably fainter. Although I've found it by scanning 8° south and slightly east of M44, it is also about a degree west of Alpha Cancri, which is the (relatively) bright star above Hydra's head.

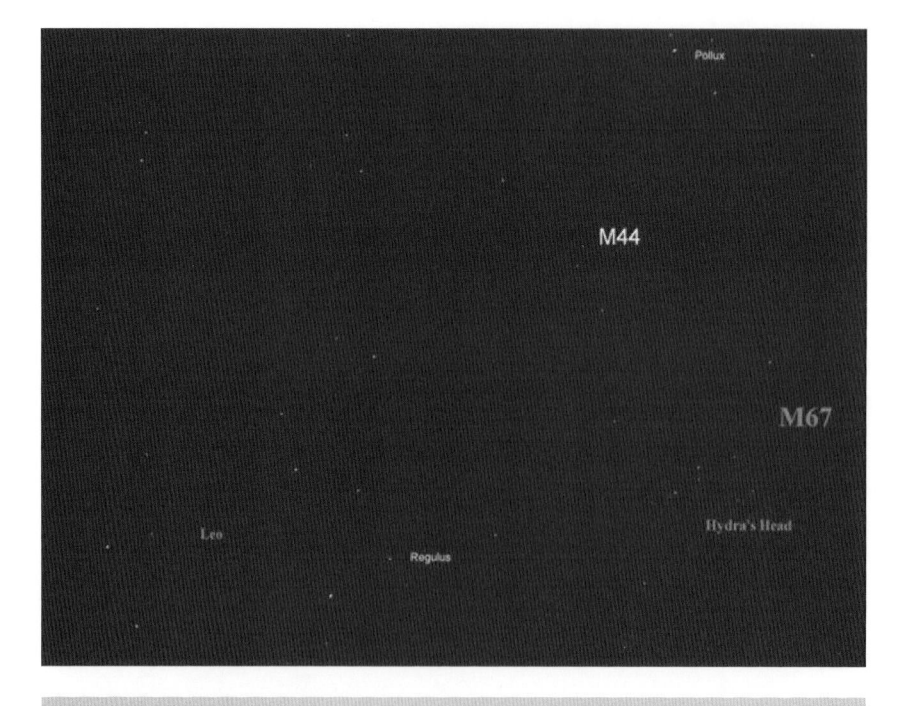

Fig. 5.55 How to find M67

Object Type	Open star cluster
Declination	11° 49 min
Right Ascension	8 h 50.4 min
Magnitude	6.1

What It Looks Like

M67 is a star cluster but it is much more sparse than the nearby Beehive (M44). However, use of a 127 mm Maksutov, armed with the usual set-up of a 32 mm Plössl eyepiece and focal reducer bring out some of the brighter stars, although the full 500 are far from visible, more like 50–80 under average cloudless conditions.

Charles Messier's Original Notes

6 April 1780
(RA: 08 h 36 m 28 s, 129d 06' 57"; Dec: +12d 36' 38"; Diameter: –)

Cluster of small stars with nebulosity, below the southern claw of the crab [Cancer]. The position determined from the star α [Cancri].

How to View

Although M67 is not a large open star cluster, it is best viewed with low magnification, as not to spoil its appearance by spreading the stars out too far. It benefits from a large aperture but not so dramatically as the Beehive does. It can be spotted with 70 mm binoculars (possibly smaller instruments on a good night from a dark site).

Photographic Details

This was taken using a 110 mm/F7 refractor and SBIG ST2000xcm camera cooled to −25°. Fifteen images, each of 1 min duration were taken and stacked and processed using *Nebulosity* (Stark Software) (Fig. 5.56).

Figure 5.57 shows a more realistic representation of M67 as seen through the usual set-up. Some of the brighter stars are lost, even though the overall shape and impression of it remains.

Fig. 5.56 M67 by Anthony Glover

M68

M68 is a globular star cluster in Hydra. It is rather rare in that its position is somewhat away from the galactic plane. It was discovered by Charles Messier in 1780 and is 33,300 light years away. It is 11 arcminutes across. Like many globular star clusters in Sagittarius, it has a low southern declination, so is never seen at its best from the United States, United Kingdom and Canada. Its brighter members are of magnitude 12.6.

Where to Find

M68 is in a sparse patch of sky and the nearest bright star is Beta Corvi. M68 is about 4° south and slightly east (Fig. 5.59).

M68

Fig. 5.57 M67 modified

Object Type	Globular star cluster
Declination	−26° 45 min
Right Ascension	12 h 39.5 min
Magnitude	8.9

What It Looks Like

M68 appears as a fuzzy patch in a 127 mm Maksutov and doesn't resolve any stars. It would probably have been invisible without the light pollution reduction filter, unless viewed from a dark site.

Charles Messier's Original Notes

9 April 1780
(RA: 12 h 27 m 38 s, 186d 54' 33"; Dec: -25d 30' 20"; Diameter: –)

Fig. 5.58 M68 (NASA/courtesy of nasaimages.org)

Nebula without stars below the Raven [Corvus] & the Hydra; it is very difficult to see with refractors; it is close to a star of 6th magnitude.

How to View

Its low southern declination makes M68 a harder target than most members of the Virgo Supercluster as viewed from mid-northern latitudes, despite being slightly brighter. Indeed it is not really a suitable target for modest instruments and really needs a large aperture telescope.

Photographic Details

Figures 5.58 and 5.60

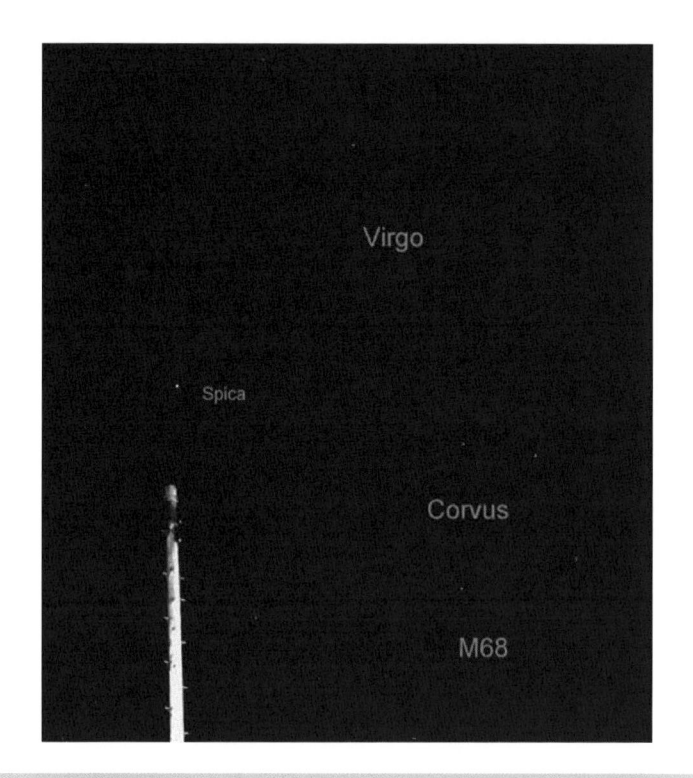

Fig. 5.59 How to find M68

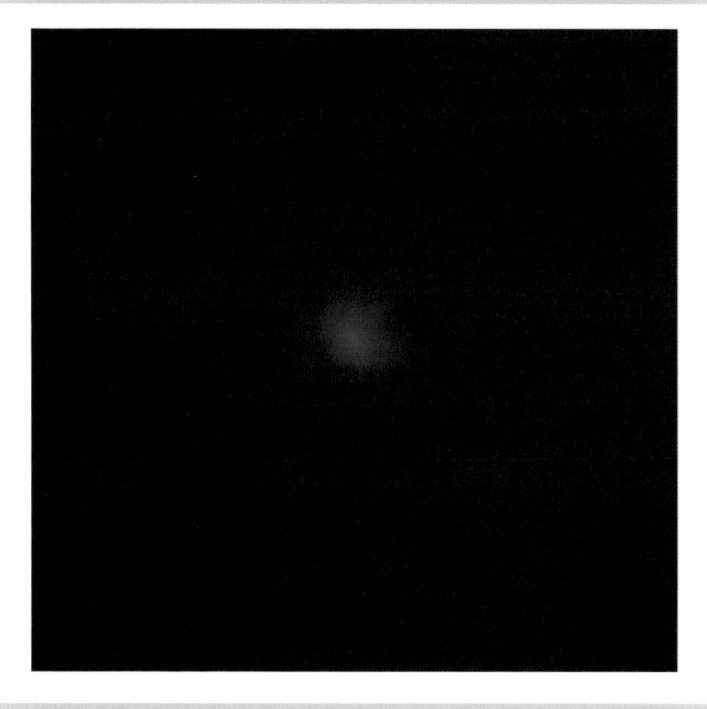

Fig. 5.60 M68 modified (NASA/courtesy of nasaimages.org)

Chapter 6

M69–M91

M69

M69 is a globular star cluster in Sagittarius and was discovered by Charles Messier in 1780. To most instruments, it appears to be about 7 arcminutes across but long exposure photographs suggest it is about 9.8 arcminutes in diameter. It is 29,700 light years away from us but only 6,200 light years away from the galactic center. None of its stars is bright enough to be seen in most amateur instruments.

Where to Find

Figure 6.1

Object Type	Globular star cluster
Declination	−32° 21 min
Right Ascension	18 h 31.4 min
Magnitude	7.6

M69 is hard to find on its own but it happens to be almost 2° due west of M70 and can be seen in the same binocular field of view.

Fig. 6.1 M69 (NASA/courtesy of nasaimages.org)

What It Looks Like

Through binoculars, M69 looks nothing special but it is easier than its magnitude would suggest as it is quite a small object. It appears as a fuzzy circular patch, just about identical to M70. A hint of central condensation is visible in a 127 mm Maksutov.

Charles Messier's Original Notes

31 August 1780
(RA: 18 h 16 m 47 s, 274d 11' 46"; Dec: −32d 31' 45"; Diameter: 0d 02')

Nebula without star in Sagittarius. Below his left arm & near it is a star of 9th magnitude; its light is very faint, and can be seen in fine weather, & the least amount of light used to illuminate the micrometer made her disappear: its position was determined by ε Sagittarii: this nebula has been observed by M. de la Caille, & reported in his Catalogue; it resembles the nucleus of a small comet.

How to View

M69 is a tough object from northern temperate latitudes and needs a large aperture telescope to see it anywhere near its best.

Photographic Details

Figures 6.1 and 6.3

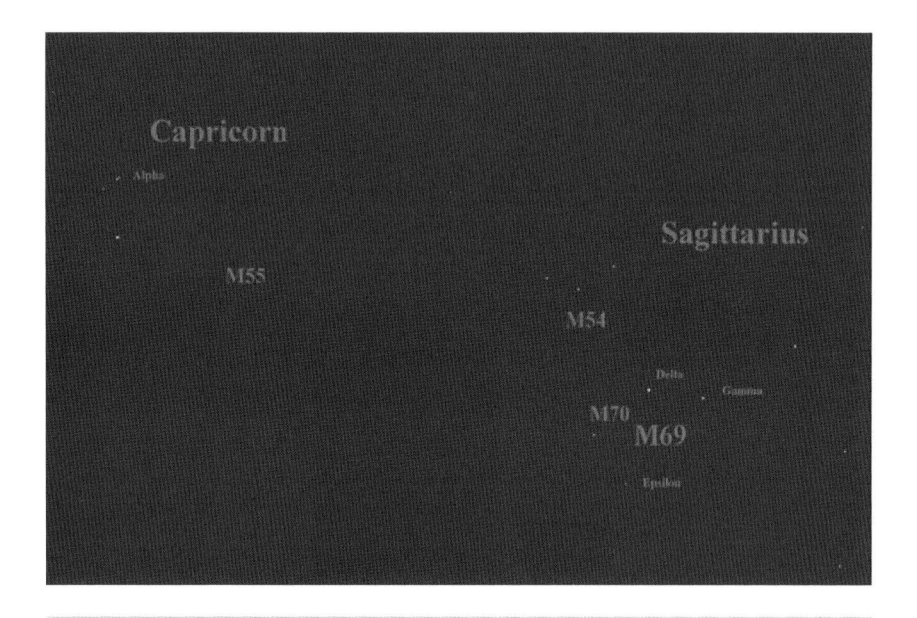

Fig. 6.2 How to find M69

Fig. 6.3 M69 (ModifiedNASA/courtesy of nasaimages.org)

M70

M70 is a globular star cluster in Sagittarius. It was discovered by Charles Messier in 1780 and at a distance of 29,300 light years and diameter of 8 arcminutes is almost a twin of its neighbour, M69. Indeed the two appear to be influenced by each other's gravity. Comet Hale-Bopp was found near M70 in 1995.

Where to Find

Figure 6.4

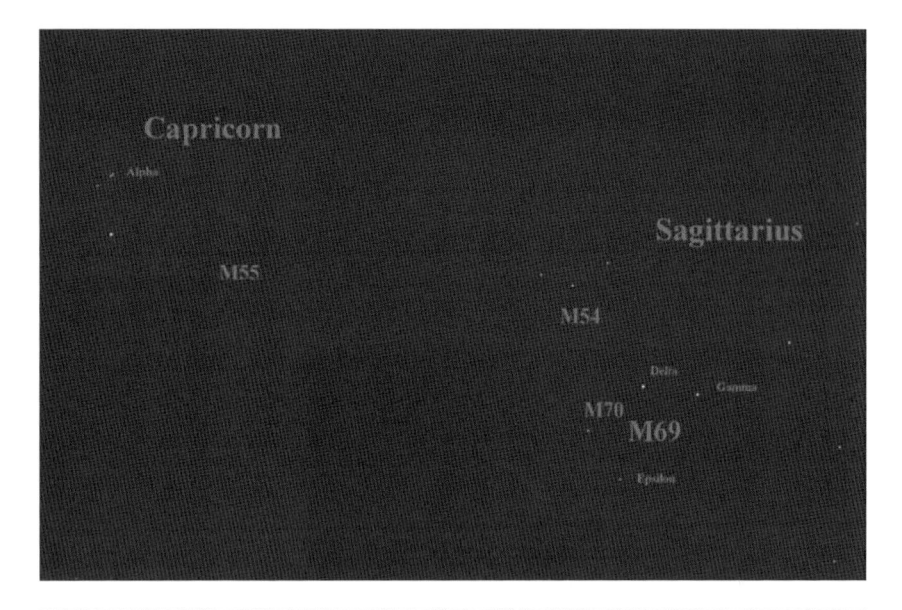

Fig. 6.4 How to find M70

Object Type	Globular star cluster
Declination	−32° 18 min
Right Ascension	18 h 43.2 min
Magnitude	8.7

M70, being one of the more southern of the Messier objects, is quite hard to find but happens to lie about 8° due south of M22. Unless you are using very high magnification, 8° due south will do the job.

What It Looks Like

Through binoculars, M70 looks nothing special but it is easier than its magnitude would suggest as it is quite a small object. It appears as a fuzzy circular patch. A hint of central condensation is visible in a 127 mm Maksutov.

Charles Messier's Original Notes

31 August 1780
(RA: 18 h 28 m 53 s, 277d 13' 16"; Dec: −32d 31' 07"; Diameter: 0d 02')

Nebula without star, near the previous one & on the same parallel: near it is a star of 9th magnitude & four small telescopic stars, almost as a single line very near each other, & are placed over the nebula as seen in a telescope that reverses [the image]; the nebula is determined from the same star ε Sagittarii.

How to View

M70 is a tough object from northern temperate latitudes and needs a large aperture telescope to see it anywhere near its best.

Photographic Details

Figures 6.5 and 6.6

Fig. 6.5 M70 (NASA/courtesy of nasaimages.org)

Fig. 6.6 M70 modified (NASA/courtesy of nasaimages.org)

M71

M71 is a globular star cluster in Sagitta. It was first discovered by Philippe Loys de Chéseaux (date uncertain). It is about 7 arcminutes in diameter but there are some surrounding stars out to 24 arcminutes but it is not certain whether they are members or not. It is 13,000 light years away and there was some doubt as to whether it is a true globular star cluster but it is now recognized as a loose one.

Where to Find

M71 lies within the faint constellation of Sagitta. Quite frankly, I've only ever seen Sagitta with the unaided eye from a dark site. Fortunately, it lies about 10° north of the first magnitude star Altair in Aquila (the southern marker of the "Summer Triangle"), which makes it easier to find (Fig. 6.8).

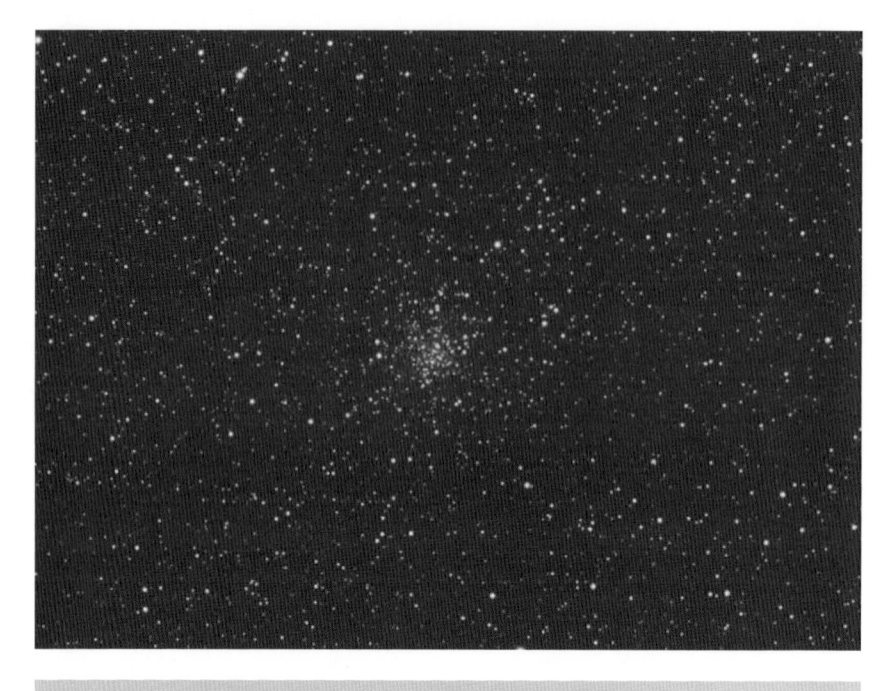

Fig. 6.7 M71 by Anthony Glover

Object Type	Globular star cluster
Declination	18° 47 min
Right Ascension	19 h 53.8 min
Magnitude	7.2

What It Looks Like

M71, through modest instruments, is more one of those objects to cross of the list of things to see before you die, rather than an aesthetically pleasing sight. Through any small instrument, provided that the transparency is good enough to see it at all, it appears as a roughly circular ghostly blob. Theoretically, the central condensation should be obvious but I noticed no difference in brightness between the visible parts, suggesting that I could only see the core. The Maksutov just spread the light too much to make it barely visible and the best view was with the finderscope.

Charles Messier's Original Notes

4 October 1780
(RA: 19 h 43 m 57 s, 295d 59' 09"; Dec: +18d 13' 00"; Diameter: 0d 03.5')
M. Méchain: (RA: 296d 00' 04"; Dec: +18d 14' 21")

Nebula discovered by M. Méchain on 28 June 1780, between the stars γ and δ of the Arrow [γ and δ Sagittarii]. On 4 October following, M. Messier has looked for [it]: its light is very faint & contains no star; the least light [to illuminate the micrometer?] makes it disappear. It is placed about 4° below that which M. Messier discovered in the Fox [Vulpecula]. See No. 27 [M27]. He reported it on the chart of the Comet of 1779.

How to View

As even the central part is much larger than any planet, low magnification, coupled with as large an aperture as possible is the best. It is unlikely to appear anything other than a faint blob in anything smaller than 200 mm.

Photographic Details

M71 was taken using a Skywatcher ED100, semi-APO, SXV H9, with an infrared blocking filter. It consisted of 16 1 min exposures (Figs. 6.7 and 6.9).

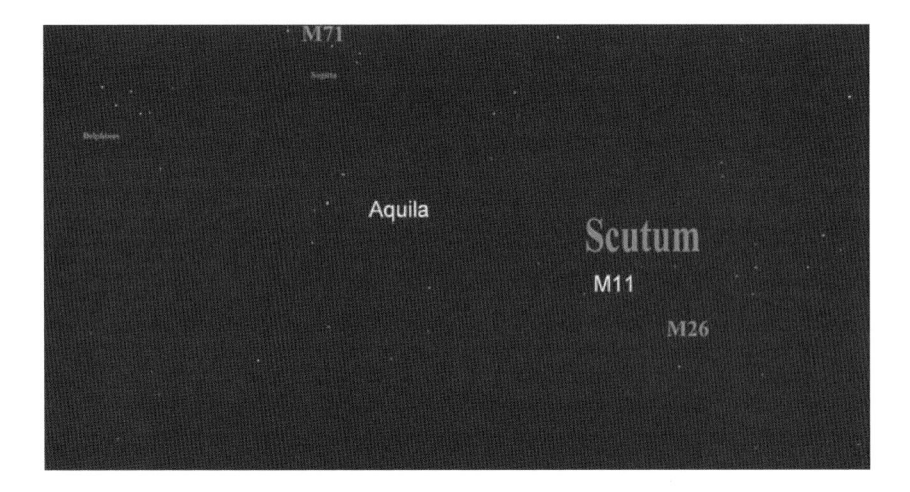

Fig. 6.8 How to find M71

Fig. 6.9 M71 modified

M72

M72 is a globular star cluster in Aquarius. It was discovered by Pierre Méchain in 1780. Its apparent size of 6.6 arcminutes and its low brightness are caused by its distance of 55,400 light years, one of the furthest globular star clusters from us and the galactic center. Like M71, it is a fairly loose cluster. Its brightest star is about magnitude 14.2.

Where to Find

Although M72 is in Aquarius, the nearest bright star is Alpha Capricorni. It is a tough object to find, as it is invisible in finderscopes. It is near M73 (Fig. 6.10).

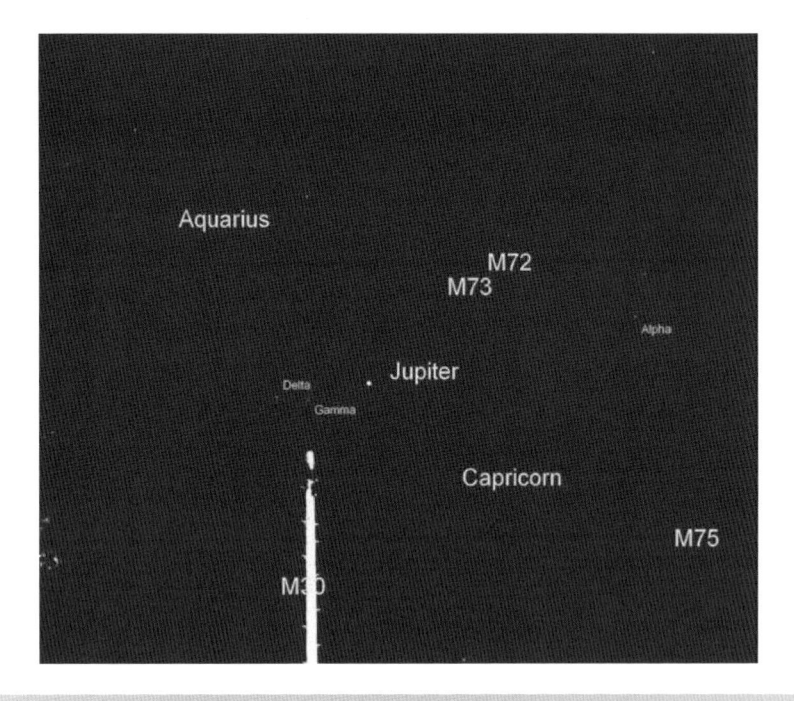

Fig. 6.10 How to find M72

Object Type	Globular star cluster
Declination	−12° 32 min
Right Ascension	20 h 53.5 min
Magnitude	9.3

What It Looks Like

M72 can be seen as a misty patch to 70 mm binoculars on a good night but through a 127 mm Maksutov takes on a rather attractive ghostly appearance but is not easily recognizable as a globular star cluster.

Charles Messier's Original Notes

4 October 1780
(RA: 20 h 41 m 23 s, 310d 20' 49"; Dec: −13d 20' 51"; Diameter: 0d 02')
M. Méchain: (RA: 310d 21' 10"; Dec: −13d 21' 24")

Nebula seen by M. Méchain on the night of 29–30 August 1780, over the neck of Capricorn. M. Messier has looked for it on 4th & 5th October following: its light is faint, as with the previous [nebula]; close to her is a small telescopic star: its position was determined by the star ν Aquarii, of 5th magnitude.

How to View

M72 is easier than the nearby M75 from the northern hemisphere because of a more favourable declination. Although a 127 mm Maksutov brings it out more clearly, especially when armed with a light pollution reduction filter, it really needs a large aperture instrument to do it justice.

Photographic Details

Figures 6.11 and 6.12

Fig. 6.11 M72 (NASA/courtesy of nasaimages.org)

Fig. 6.12 M72 modified (NASA/courtesy of nasaimages.org)

M73

M73 is one of the oddities of the Messier Catalogue, discovered in 1780 by Charles Messier himself. It appears to be a chance alignment of four stars from 10th to 12th magnitude, with the nearest being 880 light years away and the furthest 2,500 light years away. However, there is some disagreement about the distances of these stars. It is also possible that some of the field stars surrounding this group may be physically associated with some of the members of the asterism. Messier noted some nebulousity but it hasn't been detected in modern observations.

Where to Find

Although M73 is in Aquarius, the nearest bright star is Alpha Capricornii. It is a tough object to find, as it is invisible in finderscopes. It is near M72 (Fig. 6.13).

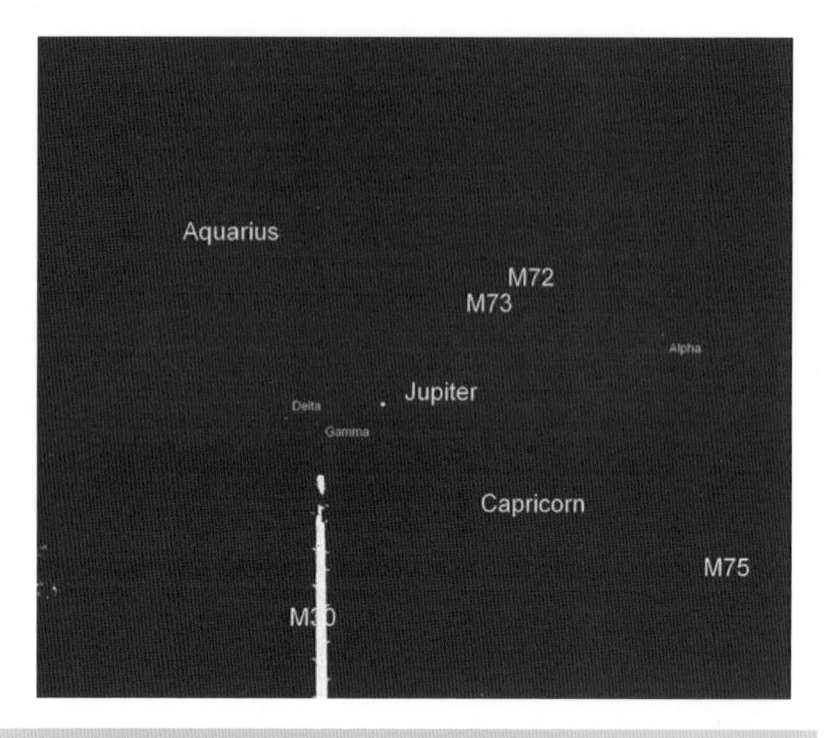

Fig. 6.13 How to find M73

Object Type	Asterism
Declination	−12° 38 min
Right Ascension	20 h 58.9 min
Magnitude	9

What It Looks Like

M73 appears as a sparse open star cluster through 70 mm binoculars, as there are a few field stars in the line of sight. In a 127 mm Maksutov, larger magnification is used, so it appears even more sparse. There was no sign of any nebulousity, as noted by Messier.

Charles Messier's Original Notes

4 and 5 October 1780
(RA: 20 h 46 m 52 s, 311d 43′ 04″; Dec: −13d 28′ 40″; Diameter: –)

M73

Clusters of three or four small stars, which resembles a nebula at first sight, contains some nebulosity: this cluster is located along the parallel of the previous nebula: its position was determined from the same star of v Aquarii.

How to View

Although a large aperture instrument brings out more field stars surrounding M73, it is important that a low magnification is used. Unless you have a short focal length telescope, large binoculars are usually better.

Photographic Details

This is a good representation of how M73 appears in a 127 mm Maksutov, so needs no modification (Fig. 6.14).

Fig. 6.14 M73 (NASA/courtesy of nasaimages.org)

M74

M74 is a spiral galaxy in Pisces. It measures 10.2×9.5 arcminutes and was discovered by Pierre Méchain in 1780. It is 35 million light years away, suggesting a size of about 95,000 light years across, similar to our Milky Way.

Where to Find

Object Type	Spiral galaxy
Declination	15° 47 min
Right Ascension	1 h 36.7 min
Magnitude	9.4

If you look in a star atlas you will notice that M74 is in the constellation of Pisces. However, the best way to find it is to use Aries. Use the two brightest stars (Alpha and Beta) as pointers, south and slightly west and you will find Eta Piscium. If you use a low power eyepiece, you will see M74 in the same field of view, as it is less than a degree away from the star (Fig. 6.16).

Fig. 6.15 M74 by Anthony Glover

What It Looks Like

M74 is not visible in small instruments. In my Maksutov, it is very reminiscent of M32, a satellite galaxy of M31. It appears like a galaxy, with a central condensation and fainter, outlying regions but needs a larger aperture than my Maksutov to resolve the spiral structure.

Charles Messier's Original Notes

(RA: 01 h 24 m 57 s, 21d 14' 09"; Dec: +14d 39' 35"; Diameter: –)

Nebula without star, near the star ε Piscium, seen by M. Méchain in late September 1780, and he reports: "This nebula contains no stars; it is quite broad, very obscure, extremely difficult to observe; we can better determine it more precisely in fine, frost [conditions]." M. Messier has sought for and has found it, as described by M. Méchain: it was compared directly to the star ε Piscium.

How to View

Although M74 is visible in my Maksutov on a good night, you need an instrument with at least 200 mm aperture to see it at its best. If you find the Pinwheel (M33) a challenge on a particular night, don't bother with M74. As it is quite large, it is best to avoid high magnification unless you have a very large instrument.

Photographic Details

Figure 6.15 is a close approximation to how M74 looks like through my Maksutov, so needs no modified version (Fig. 6.15).

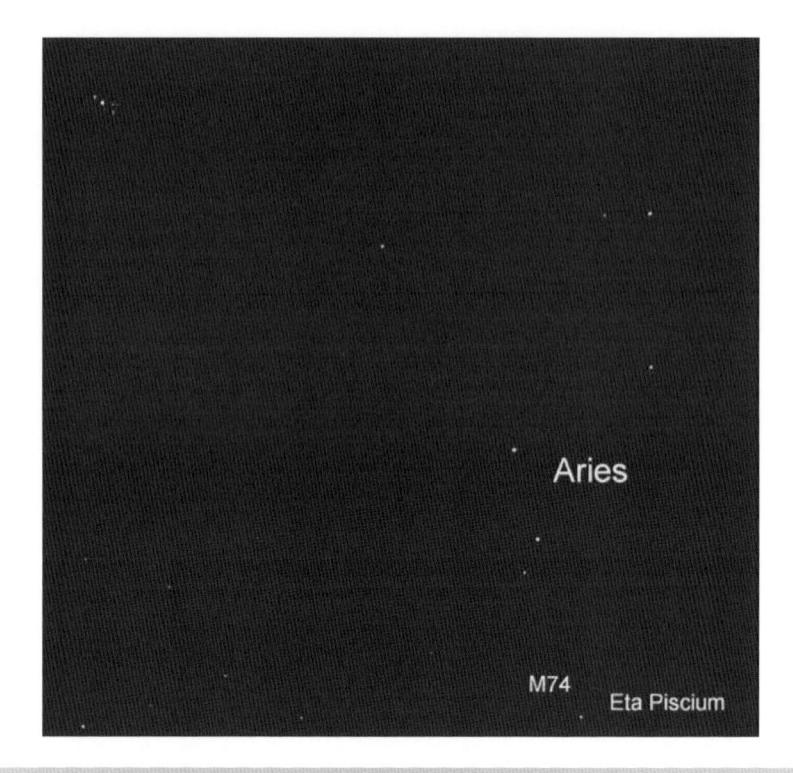

Fig. 6.16 How to find M74

M75

M75 is a globular star cluster in Sagittarius. It was discovered by Pierre Méchain in 1780. It is nearly 7 arcminutes across and lies 67,500 light years away. Due to its large distance, it cannot be resolved into stars by moderate amateur instruments.

Where to Find

M75 is in a sparse region of sky, much affected by extinction from mid-northern latitudes. The nearest brightish star is Alpha Capricornii but it is of little help (Fig. 6.17).

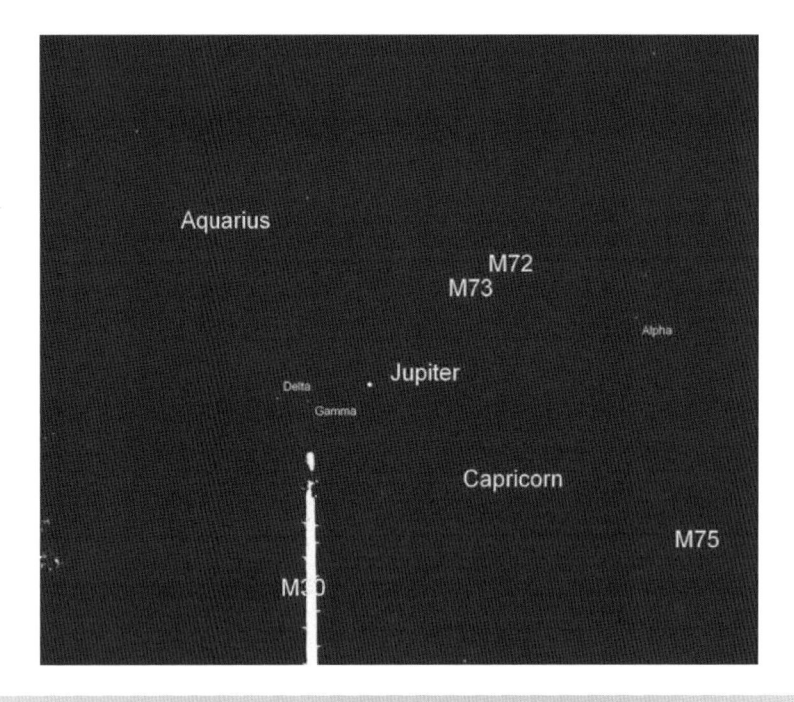

Fig. 6.17 How to find M75

Object Type	Globular star cluster
Declination	−21° 55 min
Right Ascension	20 h 6.1 min
Magnitude	8.5

What It Looks Like

M75 appears as a fuzzy patch in 70 mm binoculars but only on the clearest of nights. The view is little improved by a 127 mm Maksutov, even with a light pollution reduction filter.

Charles Messier's Original Notes

18 October 1780
(RA: 19 h 53 m 10 s, 298d 17' 24"; Dec: -22d 32' 23"; Diameter: –)
M. Méchain: (RA: 298d 17' 30"; Dec: −22d 32' 00")

Nebula without star, between Sagittarius & the head of Capricorn; seen by M. Méchain on 27th & 28th August 1780. M. Messier has looked for it on 5 October and the following 18th October, was compared to the star four Capricorni of 6th magnitude, according to Flamsteed: it seemed to M. Messier that it was composed of very small stars, containing nebulosity: M. Méchain reported it as a nebula without stars. M. Messier saw it on 5th October; but the Moon was above the horizon, & it was only the 18th of that month that he could judge her appearances and determine its place.

How to View

Although M75 can be found in modest instruments, it needs a large aperture instrument and clear night to radically improve the view.

Photographic Details

Figures 6.18 and 6.19

Fig. 6.18 M75 (NASA/courtesy of nasaimages.org)

Fig. 6.19 M75 modified (NASA/courtesy of nasaimages.org)

M76

M76 is a planetary nebula in Perseus and is also known as the Little Dumbell Nebula. It was discovered by Pierre Méchain in 1780. The brighter parts are 65 arcseconds across and its full extent is about 2.4×1.8 arcminutes across. It is surrounded by faint nebulosity nearly five arcminutes in extent, although this is unlikely to be visible in small instruments. Its distance isn't known, with estimates from 1,700 light years to as much as 15,000. In the latter case, it must be near the upper limit for this type of object. The central star is magnitude 16.6 and at 60,000° is very hot for a white dwarf. Had the progenitor star been slightly larger, it is probable that we would be seeing a supernova remnant instead of a planetary nebula.

It is one of the faintest objects in the Messier Catalogue but the good news for northern hemisphere observers is that it is circumpolar from much of Europe and the United States and during winter will suffer from little extinction. However from 35° south of the equator, it is just about impossible.

Fig. 6.20 M76 (NASA/courtesy of nasaimages.org)

Where to Find

Although M76 technically belongs to Perseus, it is actually as close to the main asterism of Cassiopeia. It is in a patch of sky devoid of 3rd magnitude stars and borders onto the Milky Way background. Its faintness also makes it impossible to see in even good finderscopes, so it is best to wait for a perfect night before viewing (Fig. 6.21).

Object Type	Planetary nebula
Declination	51° 34 min
Right Ascension	1 h 42.4 min
Magnitude	10.1

What It Looks Like

M76 is a very tough object for medium aperture telescopes and, indeed, needed optimum conditions for it to be visible at all from a suburban site. Through a 127 mm Maksutov, it looks more like the Ring Nebula (M57) but this is hardly a surprise, as it is very near the limit of that type of telescope.

M76

Charles Messier's Original

21 October 1780
(RA: 01 h 28 m 43 s, 22d 10' 47"; Dec: +50d 28' 48"; Diameter: 0d 02')
M. Méchain: (RA: 22d 10' 26"; Dec: +50d 28' 12")

Nebula to the right foot of Andromeda, seen by M. Méchain on 5th September 1780, & he reports: "This nebula contains no stars; it is small & faint." On the following 21st October, M. Messier sought it with the his achromatic telescope, & it seemed that it was composed of very small stars, which contained nebulosity, & that the least light used to illuminate the micrometer makes it [the nebula] disappear: the position was determined from the star φ Andromedae of 4th magnitude.

How to View

Being one of the hardest objects in the list, even when well placed, it needs a dark site, exceptional viewing conditions and a large aperture to do it justice. Otherwise, it is merely an object to tick off the list of things to see before you die.

Photographic Details

Figures 6.20 and 6.22

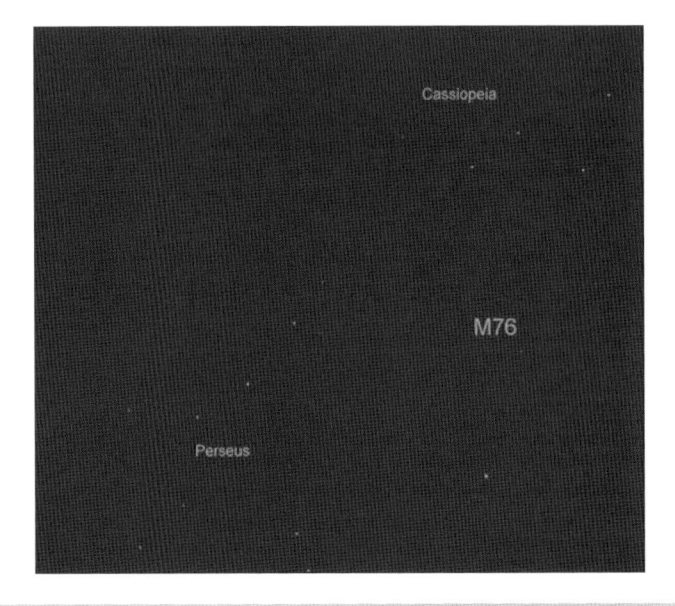

Fig. 6.21 How to find M76

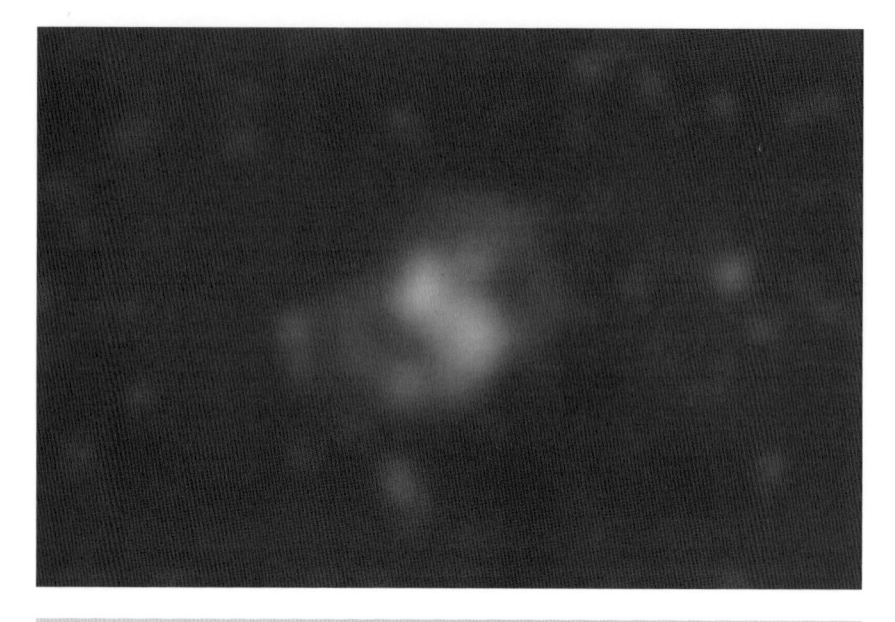

Fig. 6.22 M76 modified (NASA/courtesy of nasaimages.org)

M77

M77 is a spiral galaxy in Cetus very close to the celestial equator. It was discovered by Pierre Méchain in 1780. It has an apparent size of 7×6 arcminutes and is about 60 million light years away (about the same as the Virgo Supercluster). This makes its main dimension as much as 120,000 light years across, although outlying regions may extend as far as to make it 170,000 light years across, truly large for a spiral galaxy. M77 is known as a Seyfert Galaxy, after Carl Keenan Seyfert who discovered this type of object. It has an active galactic nucleus, which is thought to be some sort of mini quasar, that accelerates gas and stars around a central massive object (believed to be a black hole) and emits a lot of radiation at infra red and radio wavelengths. Being near the celestial equator, it is visible from most of the world, except the polar regions.

Where to Find

M77 is very near Delta Ceti, near the "head" of the constellation, with Delta being the northernmost bright star in the neck of Cetus. M77 is too faint to be visible in finderscopes but can be seen through most telescopes when Delta Ceti is in the center of the field of view (Fig. 6.24).

Fig. 6.23 M77 (NASA/courtesy of nasaimages.org)

Object Type	Spiral galaxy
Declination	−00° 01 min
Right Ascension	2 h 42.7 min
Magnitude	8.9

What It Looks Like

M77 is a tough object for medium aperture instruments. Although it is slightly brighter than the nearby M74, its lower altitude makes it about the same difficulty. Most nights it is invisible to a 127 mm Maksutov but when it can be seen, shows a central nucleus and some outlying "fuzzy" regions. It cannot be distinguished as a spiral galaxy unless you use larger instruments.

Charles Messier's Original Notes

17 December 1780
(RA: 02 h 31 m 30 s, 37d 52' 33"; Dec: −00d 57' 43"; Diameter: −)
M. Méchain: (RA: 37d 52' 58"; Dec: −00d 57' 44")

Cluster of small stars, which contains nebulosity, in the Whale [Cetus], & on the same parallel as the star δ [Ceti] reported as being of 3rd magnitude, & that M. Messier hasn't found to be of the 5th. M. Méchain saw this cluster on 29 October 1780 in the form of a nebula.

How to View

M77 is normally invisible to medium aperture instruments but when local conditions allow, a 127 mm telescope will show it as a galaxy from suburban locations. To see its spiral structure, you need a dark site, great conditions and a large aperture instrument. A light pollution reduction filter helps bring out a bit more detail.

Photographic Details

Figures 6.23 and 6.25

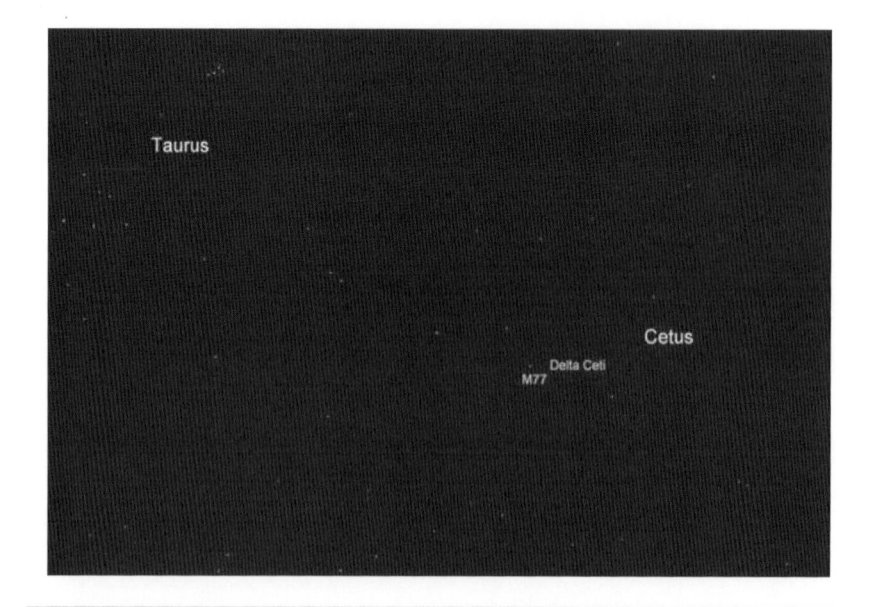

Fig. 6.24 How to find M77

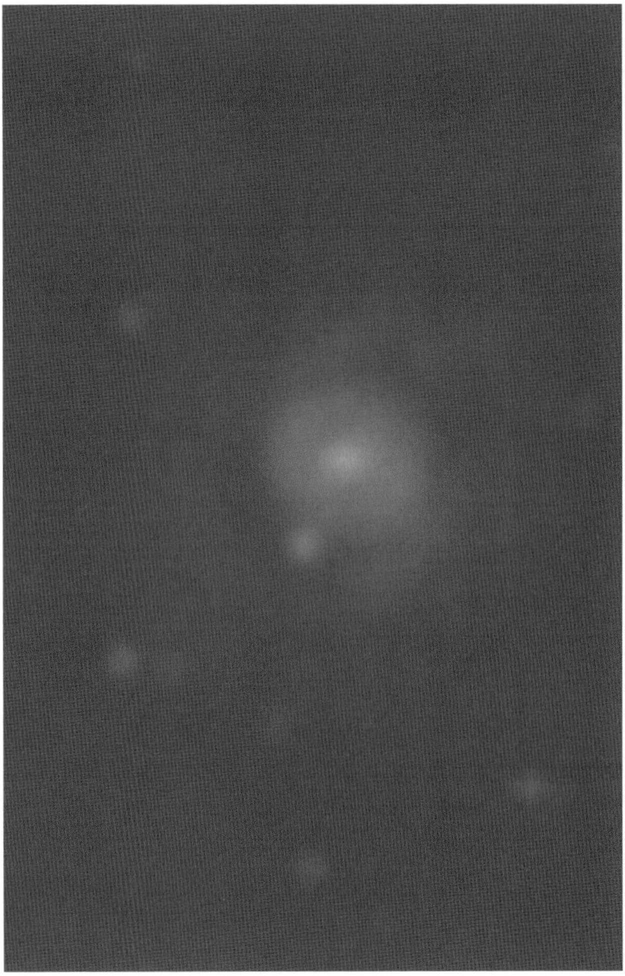

Fig. 6.25 M77 modified (NASA/courtesy of nasaimages.org)

M78

M78 is a nebula in Orion which is similar in nature to the Orion Great Nebula and may, indeed, even be associated with it. It was discovered in 1780 by Pierre Méchain. Although, like its better known neighbor, its distance is not well established, it is believed to be about 1,600 light years away and is about 8×6 arcminutes in size. Research has shown that many newly formed stars are

Fig. 6.26 M78 by Anthony Glover

present around M78 and that many more are embedded within it. It is also possible that much of its true extent is hidden from view by dark nebulae in the foreground.

Being near the celestial equator, it is well-placed from most of the world.

Where to Find

M78 is found about 2° north of Zeta Orionis, the eastmost star in Orion's belt and less than a degree to the east. It can be found by scanning from Zeta Orionis in the direction of Betelguese. It is difficult to see in a finderscope, even on a very clear night (Fig. 6.27).

Object Type	Nebula
Declination	0° 3 min
Right Ascension	5 h 46.7 min
Magnitude	8.3

M78

What It Looks Like

M78 appears as an irregular misty patch, suggesting that it is a nebula, rather than a galaxy. My fist view through my 127 mm Maksutov-Cassegrain was, frankly, a disappointment. A revisit, using the same set-up but with a light pollution reduction filter was a huge improvement and it was nice to see what appeared like a star cluster in the vicinity.

Charles Messier's Original Notes

17 December 1780
(RA: 05 h 35 m 34 s, 83d 53' 35"; Dec: -00d 01' 23"; Diameter: 0d 03')
M. Méchain: (RA: 83d 53' 02"; Dec: -00d 00' 31")

Star cluster, with lots of nebulosity in Orion and on the same parallel as the star δ [Orionis–a.k.a. Mintaka] on the Belt, which was used to determine its position; and followed the star cluster over the hour at 3d and 41', and the cluster is above the star by 27' 7". M. Méchain saw this cluster at the beginning of 1780, and reported as follows: "On the left side of Orion, 2–3 min in diameter, one sees two fairly bright nuclei, surrounded by nebulosity."

How to View

M78 is a tough object that is not visible in finderscopes and small aperture telescopes, so the larger aperture you can lay your hands on, the better. A light pollution reduction filter is a great help from suburban skies.

Photographic Details

Figuers 6.26 and 6.28

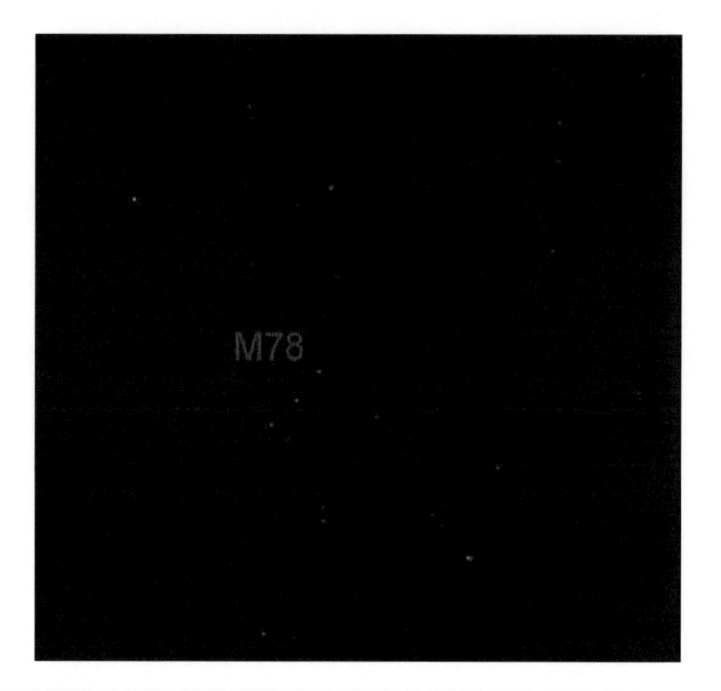

Fig. 6.27 How to find M78

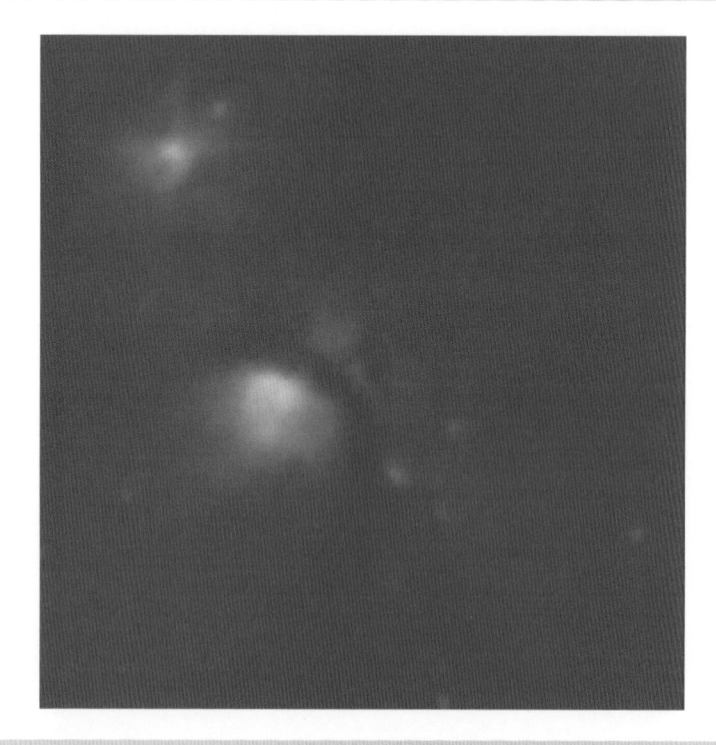

Fig. 6.28 M78 modified (NASA/courtesy of nasaimages.org)

M79

M79 is a globular star cluster in Lepus. It was discovered by Pierre Méchain in 1780 and is a strange place to find a globular star cluster, being about 42,000 light years from us and 60,000 miles from the galactic center. It is also very inclined to the galactic plane and one theory suggests that it is extra-galactic in origin and was captured from the Canis Major Dwarf Galaxy, which is merging with our Milky Way.

Where to Find

M79 is a tough object to find for several reasons. Firstly, it never rises far above the horizon from the United Kingdom and not much further from the Northern USA. It is hosted by the faint constellation of Lepus, which is difficult to see from suburban skies. If Alpha and Beta Lepi are visible, scan from Alpha to Beta and continue for about the same distance to find M79. It is not visible in finderscopes (Fig. 6.30).

Fig. 6.29 M79 (Courtesy of Joe Billington, Manchester Astronomical Society)

Object Type	Globular star cluster
Declination	−24° 33 min
Right Ascension	5 h 24.5 min
Magnitude	7.7

What It Looks Like

M79 is very faint and best thought of something to cross off of the list of things to see before you die. It had a more wispy appearance than I would have expected in my 127 mm Maksutov-Cassegrain but I'd say it looks like either a globular star cluster or elliptical galaxy. Although I've spotted it from the UK in binoculars, it was far from convincing. However, from near Chicago, it appeared much more promising when suitably placed near the meridian, although it was not obvious quite what sort of object it was.

Fig. 6.30 How to find M79

Charles Messier's Original Notes

17 December 1780
(RA: 05 h 15 m 16 s, 78d 49' 02"; Dec: −24d 42' 57"; Diameter: −)
M. Méchain: (RA: 78d 47' 10"; Dec: −24d 44' 46")

Nebula without stars located below the Hare [Lepus], & on a parallel as a star of 6th magnitude: seen by M. Méchain on 26th October 1780. M. Messier looked for it the following 17th December: this nebula is beautiful; the center bright, with just diffuse nebulosity; its position was determined from the star ε Leporis, of 4th magnitude.

How to View

M79 is one of the reasons that deep sky enthusiasts like large aperture instruments.

Photographic Details

Figure 6.31 is a good representation of how M79 appears in a 127 mm Maksutov.

Fig. 6.31 M79 modified (Courtesy of Joe Billington, Manchester Astronomical Society)

M80

M80 is a globular star cluster in Scorpius and was discovered by Charles Messier in 1781. It is 10 arcminutes across and is about 32,600 light years away. It is one of the densest globular star clusters and the presence of "blue stragglers" suggests that many of its stars have undergone mergers and collisions. A nova was discovered on May 21st 1860 in M80 and its peak brightness was 7.0, brighter than the whole cluster. It is not completely certain whether this nova was a cluster member or it was just a line of sight effect.

Due to its density, M80 cannot easily be resolved into stars by modest instruments and its appearance is little different from many elliptical galaxies.

Where to Find

M80 is in a sparse area of sky but is fortunately close to Antares and M4 in Scorpius. It is roughly north west of Antares by about 4° (Fig. 6.33).

Fig. 6.32 M80 (NASA/courtesy of nasaimages.org)

M80

Object Type	Globular star cluster
Declination	−22° 59 min
Right Ascension	16 h 17.0 min
Magnitude	7.3

What It Looks Like

M80 looks like a fuzzy patch of indeterminate type through 70 mm binoculars and a 127 mm Maksutov shows it as being larger but it's true nature is somewhat indiscernible.

Charles Messier's Original Notes

4 January 1781
(RA: 16 h 04 m 00 s, 240d 59' 48"; Dec: −22d 25' 13"; Diameter: 0d 02')
M. Méchain: (RA: 241d 00' 26"; Dec: -22d 27' 58")

Nebula without star, in the Scorpion [Scorpius], between the stars *g* & δ Scorpii, compared to *g* to determine its position: this nebula is round, brilliant in the center & resembles a small nucleus of a comet, surrounded by nebulosity. M. Méchain saw it on 27 January 1781.

Fig. 6.33 How to find M80

Fig. 6.34 M80 modified (NASA/courtesy of nasaimages.org)

How to View

M80 looks like a globular star cluster or elliptical galaxy in modest instruments and really needs a large aperture instrument to see any real detail. Its low southern declination makes it difficult from mid-northern latitudes.

Photographic Details

Figures 6.32 and 6.34

M81

M81 is a spiral galaxy in Ursa Major and was discovered by Johann Elert Bode in 1774. It is 12 million light years away and is quite large at 21 × 10 arcminutes. It is only 150,000 light years away from its neighbour, M82, and these two are the

Fig. 6.35 M81 by Anthony Glover

dominant galaxies in a small group. They also appear to have had a close encounter and have warped each other, triggering bursts of star formation. From a research perspective, M81 appears to have less than its fair share of dark matter. A supernova was discovered in M81 in 1993.

At least seven observers have seen M81 without optical aid but this is not a realistic aim from medium light polluted skies. Indeed, I haven't included it in my list of easy objects for 50 mm aperture binoculars and, even in larger binoculars, it only appears as a fuzzy patch where I happen to know where M81 should be. On exceptionally clear nights, notably in January 2006, I was able to see the spiral structure in my 15×70 binoculars.

Where to Find

M81 is in a difficult part of sky. It is not near any bright object and is best found by following a line between Gamma and Alpha Ursae Majoris for about the same distance. On a good night from a dark site, M81 is visible in a large finderscope (Fig. 6.36).

Object Type	Spiral galaxy
Declination	69° 4 min
Right Ascension	9 h 55.6 min
Magnitude	6.9

What It Looks Like

M81 is one of the brighter galaxies in the list. It is circumpolar from most of the northern hemisphere but is best placed in spring. On an exceptional night, it is possible to see the nucleus and surrounding disc in 70 mm binoculars but it needs something a bit larger, such as a 127 mm instrument to see a hint of the spiral structure. However, on an average night it is invisible to 70 mm binoculars.

Charles Messier's Original Notes

9 February 1781
(RA: 09 h 37 m 51 s, 144d 27′ 44″; Dec: +70d 07′ 24″; Diameter: –)
M. Méchain: (RA: 144d 27′ 00″; Dec: +70d 04′ 00″; Diameter: –)

Nebula near the ear of the Great Bear [Ursa Major], on the parallel of the star *d* of 4th to 5th magnitude: its position was determined via this star. This nebula is somewhat oval, the center clear, & we see it very well with a regular telescope of 3 ft & a half. It was discovered in Berlin by M. Bode, 31st December 1774, & by M. Méchain, in August 1779.

How to View

At the risk of this being a cliche, M81 is best viewed with a large aperture instrument on a clear night from a dark location. However, it is one of the brighter galaxies and can certainly be detected in 70 mm instruments, although larger instruments are needed to distinguish it from an elliptical galaxy. A light pollution reduction filter is a great help.

Photographic Details

M81 was taken using 5×10 min exposures with an 110 mm/F7 refractor, SBIG St2000×cm, dark subtracted, stacked and processed in *Nebulosity* (Fig. 6.35).

Figure 6.37 was taken using a 10″ F5 Newtonian, Starlight Xpress H×916, guided with a 80ED refractor with a modifed Toucam. It consisted of 120 5 min exposure luminance frames, ten 5 min exposures of hydrogen alpha frames (binned 2×2) and 15 5 min exposures of RGB (Fig. 6.38).

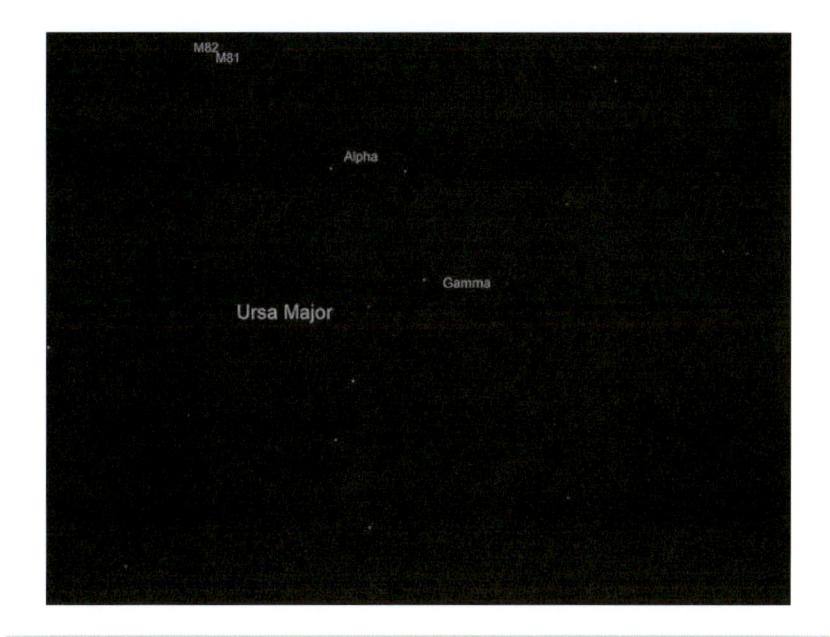

Fig. 6.36 How to find M81

Fig. 6.37 M81 by Mike Deegan

272   6   M69–M91

Fig. 6.38   M81 modified

M82

M82 is an irregular galaxy in Ursa Major. It was discovered by Johann Elert Bode in 1774. It is 12 million light years away and is 9×4 arcminutes across. It is very close to M81, being about 150,000 light years away and it is likely that it is gravitationally interacting with its neighbor. It has a very distorted shape. M82 is radiating a lot of energy at radio and infra red wavelengths.

M82 is circumpolar from all parts of the United States and Europe but is best seen in spring.

Where to Find

Please refer to Figure 6.35, as it is very close to M81.

Object Type	Irregular galaxy
Declination	69° 41 min
Right Ascension	9 h 55.8 min
Magnitude	8.4

M82 273

What It Looks Like

M82 appears as a fuzzy patch to 70 mm binoculars on a good night. On an exceptional night, it looks like a cigar-shaped object. Although its magnitude of 8.4 would suggest it could be a tough object, it is slightly easier than the nearby M81. Most of its light is concentrated in a very thin band and through a 127 mm Maksutov looks rather like many of the photographs but without the colors, as a twisted, distorted line, with some faint fuzziness surrounding it.

Charles Messier's Original Notes

9 February 1781
(RA: 09 h 37 m 57 s, 144d 29' 22"; Dec: +70d 44' 27"; Diameter: –)
M. Méchain: (RA: 144d 28' 13"; Dec: +70d 43' 05"; Diameter: –)

Nebula without star, near the previous one; with both appearing at the same time within the bezel of the telescope, this one is less distinct than the previous, with a faint & elongated light: at its end is a telescopic star. Viewed in Berlin, by M. Bode, on 31 December 1774, & by M. Méchain in August 1779.

How to View

Due to their proximity, it is almost inconceivable that anyone would look at M82, without observing M81 at the same time. However, M82 is a slightly easier galaxy and on borderline nights, M81's nucleus may be mistaken for a star, as the outer regions may not be visible. A light pollution reduction filter is a great help.

Photographic Details

Figure 6.39 was taken using a 10" F5 Newtonian, Starlight Xpress H×916, guided with a 80ED refractor with a modifed Toucam. It consisted of ten 5 min exposure luminance frames, ten 5 min exposures of hydrogen alpha frames (binned 2×2) and 12 15 min exposures of RGB.
 Figure 6.40

Fig. 6.39 M82 by Mike Deegan

Fig. 6.40 M82 modified

M83

M83 is a spiral galaxy in southern Hydra. It was discovered in 1752 by Abbe Nicholas Louis de la Caille. It is the southernmost galaxy in the Messier Catalogue and he noted it as being difficult from Paris (49° north). It is quite a tough object from most European countries and Canada. It is 11×10 arcminutes across and about 15 million light years away. It appears to be in a group of galaxies with Centaurus A, a galaxy rich in radio wavelengths and is too far south to be seen from France.

Where to Find

M83 lies on the Hydra/Centaurus border, highly unsuitable from southern England. Gamma Hydri is the brightest nearby star. M83 is also about 13° south of Spica and slightly to the east by about a degree. Some of the faint stars at the head of Centaurus are visible in a 50 mm finderscope and M83 lies almost on a line between them and Gamma Hydri (Fig. 6.42).

Fig. 6.41 M83 (Courtesy of Ray Grover, Manchester Astronomical Society)

Object Type	Spiral galaxy
Declination	−29° 52 min
Right Ascension	13 h 37.0 min
Magnitude	7.6

What It Looks Like

M83 is beyond the reach of 70 mm binoculars from southern England but the slightly larger aperture of the Maksutov shows a fuzzy patch with spiral arms blinking in and out of view. One spiral arm is particularly prominent and the overall view is reminiscent of many sketches of "spiral nebulae" before their true nature and distance was known.

Charles Messier's Original Notes

17 February 1781
(RA: 13 h 24 m 33 s, 201d 08′ 13″; Dec: −28d 42′ 27″; Diameter: −)

Nebula without star, near the head of Centaurus: it appears as a faint and even light but is so difficult to see with the telescope that the lesser light required to illuminate the micrometer made her disappear. Only with the utmost concentration can it be seen: it forms a triangle with two stars estimated to be of 6th & 7th magnitude: [the position was] determined by the stars *i, k, h*, in the head of Centaurus: M. de la Caille has already determined the nebula. See the end of this Catalogue.

How to View

From northern temperate latitudes, a large aperture is a definite help but if you get the chance to travel to southern Europe or further south, a medium aperture will suffice.

Photographic Details

Figure 6.41 is a good representation of how M83 appears in a 127 mm Maksutov.

Fig. 6.42 How to find M83

M84

M84 is a lenticular galaxy in Virgo and is one of the members of the Virgo Supercluster. It is 60 million light years away, like most of the cluster members and is 5 arcminutes across. It was discovered by Charles Messier in 1781. Three supernovae have been found in M84 in 1957, 1980 and 1991. It was also found to have a supermassive central object of 300 million solar masses, believed to be a black hole.

Where to Find

With modest equipment, M84 is difficult to find on its own, so it is best to find M87 first (see M87: Where to find) and move just over a degree in the west north west direction. It is very close to M86 (Fig. 6.43).

Fig. 6.43 How to find M84

Object Type	Lenticular galaxy
Declination	12° 53 min
Right Ascension	12 h 25.1 min
Magnitude	9.1

What It Looks Like

M84 appears as a fuzzy patch in a 127 mm Maksutov in the same wide field of view as M87. It is also even closer to M86, which appears very similar. It is to faint to discern any obvious structure in modest instruments.

Charles Messier's Original Notes

18 March 1781
(RA: 12 h 14 m 01 s, 183d30' 21"; Dec: +14d07' 01"; Diameter: –)

M84

Fig. 6.44 M84 by Anthony Glover

Nebula without star, in Virgo, the centre is a bit bright, surrounded by slight nebulosity: its brightness and appearance resemble that of those in this Catalogue, No's. 59 & 60 [M59 and M60].

How to View

Although a light pollution reduction filter will help you to detect M84 with modest instruments, it is more an object to cross of the list, unless you are able to view it through a large aperture telescope.

Photographic Details

Figure 6.44 shows M84 and its companions in the Virgo Supercluster. It was taken using a 80 mm/F6 refractor and SBIG ST2000×cm camera cooled to −25°. Six images, each of 10 min duration were taken and stacked and processed using *Nebulosity* (Stark Software). M84 is the second largest galaxy, near the top.

M84 appears as an elliptical galaxy in a 127 mm Maksutov, as shown in Fig. 6.45.

Fig. 6.45 M84 modified

M85

M85 is a lenticular galaxy in Coma Berenices. It is a member of the Virgo Supercluster (being the most northern) and shares its common distance of 60 million light years. It was discovered by Pierre Méchain in 1781 and caused Charles Messier to search the whole area and discover many of the objects we now know to be members of this supercluster. It is 7.1 × 5.2 arcminutes across.

Where to Find

M85 is in sparse part of sky. It is about 10° south of the open cluster Melotte 111 and too faint to see in a finderscope. It needs a really clear night, luck and patience (Fig. 6.47).

Object Type	Lenticular galaxy
Declination	18° 11 min
Right Ascension	12 h 25.4 min
Magnitude	9.1

M85

Fig. 6.46 M85 (NASA/courtesy of nasaimages.org)

What It Looks Like

Through a 127 mm Maksutov, the lenticular nature of M85 is not obvious. It looks like a spherical galaxy with a pronounced nucleus. However, it is not a difficult object to spot.

Charles Messier's Original Notes

18 March 1781
(RA: 12 h 14 m 21 s, 183d35' 21"; Dec: +19d 24' 26"; Diameter: –)
M. Méchain: (RA: 183d 35' 45"; Dec: +19d 23' 00"; Diameter: –)

Nebula without star, above & near the ear of the Virgin [Virgo], between two stars of the hair of Berenice [Coma Berenices], No's. 11 & 14 of the Catalogue of Flamsteed: this nebula is very faint. M. Méchain had its position determined on 4 March 1781.

How to View

A light pollution reduction filter helps show some superficial structure in a small telescope but it really needs a large aperture telescope to do it justice.

Photographic Details

Figures 6.46 and 6.48

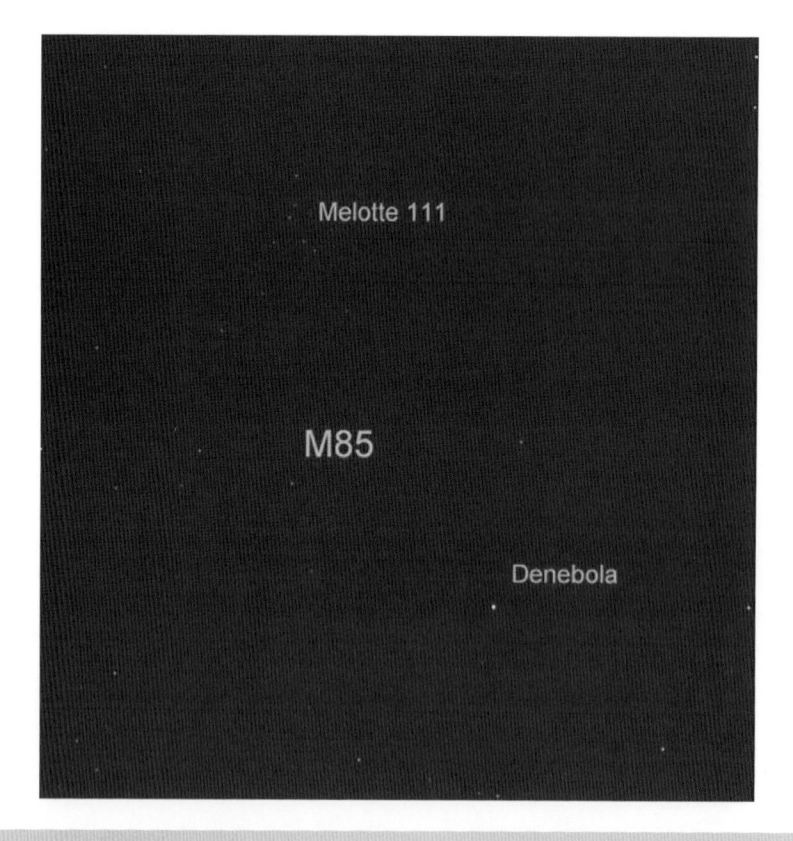

Fig. 6.47 How to find M85

Fig. 6.48 M85 modified (NASA/courtesy of nasaimages.org)

M86

M86 is a lenticular galaxy in Virgo and is a member of the Virgo Supercluster, being close to its center. It was discovered by Charles Messier in 1781. It is 7.5 × 5.5 arcminutes across and is one of the larger galaxies in the region. It is very close to M84.

Where to Find

With modest equipment, M86 is difficult to find on its own, so it is best to find M87 first (see M87: Where to find) and move just over a degree in the west north west direction. It is very close to M84.

Object Type	Lenticular galaxy
Declination	12° 57 min
Right Ascension	12 h 26.2 min
Magnitude	8.9

What It Looks Like

M86 appears as a fuzzy patch in a 127 mm Maksutov in the same wide field of view as M87. It is also even closer to M84, which appears very similar. It is too faint to discern any obvious structure in modest instrument.

Charles Messier's Original Notes

18 March 1781
(RA: 12 h 15 m 05 s, 183d 46′ 21″; Dec: +14d 09′ 52″; Diameter: –)

Nebula without star in Virgo, on the same parallel & very near the nebula above, No. 84 [M84]: and the appearances are the same, & one and the other appeared together in the same field of the telescope.

How to View

A light pollution reduction filter helps show some superficial structure in a small telescope but it really needs a large aperture telescope to do it justice.

Photographic Details

Please refer to Fig. 6.44. M86 is the largest galaxy in the diagram.
Figure 6.49 shows M86 as it looks through a 127 mm Maksutov.

Fig. 6.49 M86 modified

M87

M87 is an elliptical galaxy in Virgo. It was discovered by Charles Messier in 1781 but we now know it to be the largest and most central member of the Virgo Supercluster. It shares a common distance of 60 million light years and measures seven arcminutes across, corresponding to a linear size of 120,000 light years. Due to its elliptical (as opposed to spiral) shape, it is much more massive than either our own Milky Way or the Andromeda Galaxy (M31). Professional instruments have detected that its outer regions extend to about half a million light years. It also has a large number of globular star clusters, maybe even as many as 15,000. It is also immersed in intergalactic material and a jet of hot gas has been seen in professional instruments. It also appears to be interacting with its neighbors gravitationally.

Where to Find

M87 is one of the brighter and larger galaxies in the Virgo Supercluster. It is about 5° west and a degree north of Epsilon Virginis and M60 (which is almost as bright)

Fig. 6.50 How to find M87

lies almost in a straight line between Epsilon and M87 and is about halfway along, so M60 can be used as a "signpost" to M87 and vice-versa. M87 is also a convenient signpost for M58, M84, M86, M89 and M90 (Fig. 6.50).

Object Type	Elliptical galaxy
Declination	12° 24 min
Right Ascension	12 h 30.8 min
Magnitude	8.6

What It Looks Like

M87 appears circular in shape, as a 127 mm Maksutov cannot distinguish the difference in length of 0.4 arcminutes between semi major and semi minor axes easily. Although not spectacular in modest instruments, its size distinguishes it easily from other nearby fuzzy patches, as it is larger and brighter.

Charles Messier's Original Notes

18 March 1781
(RA: 12 h 19 m 48 s, 184d 57′ 06″; Dec: +13d 38′ 01″; Diameter: –)

Nebula without star, in Virgo, below & close enough to a star of 8th magnitude, the star having the same Right Ascension as the nebula, & its Declination of 13d 42′ 2″ north. This nebula appeared to have the same light as the two nebulae No's. 84 and 86 [M84 and M86].

How to View

Although M87 can certainly be seen in small instruments, it needs a large aperture telescope to show it properly. A light pollution reduction filter seems to make little or no difference.

Photographic Details

Figure 6.51 shows M87 and some faint companions in the Virgo Supercluster. It was taken using a 80 mm/F6 refractor and SBIG ST2000×cm camera cooled to

Fig. 6.51 M87 by Anthony Glover

−25°. Four images, each of 15 min duration were taken and stacked and processed using *Nebulosity* (Stark Software). Through a 127 mm Maksutov, M87 appears much as in the photograph but neither the faint companion galaxies nor the fainter background stars are visible.

M88

M88 is a spiral galaxy in Coma Berenices and is an outlying member of the Virgo Supercluster and shares its common distance of 60 million light years. It was discovered by Charles Messier in 1781. It is 7×4 arcminutes across, as is inclined at an angle of 30°. This corresponds to an actual size of about 120,000 light years across, somewhat larger than our own Milky Way.

Where to Find

M88 and M91 form a close pair less than a degree apart. However, even M88 is quite faint and there are no obvious guideposts. One method is to find M98 and move 3° east to M88. You can also find M91 by moving two and a half degrees north of M89. If you are unable to see any of these guidepost galaxies, M91 won't be visible and even M88 will be doubtful (Fig. 6.53).

Fig. 6.52 M88 (NASA/courtesy of nasaimages.org)

M88

Object Type	Spiral galaxy
Declination	14° 25 min
Right Ascension	12 h 32.0 min
Magnitude	9.6

What It Looks Like

Although M88 is a spiral galaxy, it appears as a small fuzzy patch of indeterminate nature in a 127 mm Maksutov on a good night. It forms a close pair with M91.

Charles Messier's Original Notes

18 March 1781
(RA: 12 h 21 m 03 s, 185d 15' 49"; Dec: +15d 37' 51"; Diameter: –)

Nebula without star, in Virgo, between two small stars & a star of 6th magnitude, which appear at the same time as the nebula in the field of the telescope. Its light is one of the faintest & resembles that reported in the Virgin [Virgo], No. 58 [M58].

How to View

M88 appears unspectacular with a modest instrument and really needs a large aperture telescope to do it justice.

Photographic Details

Figures 6.52 and 6.54

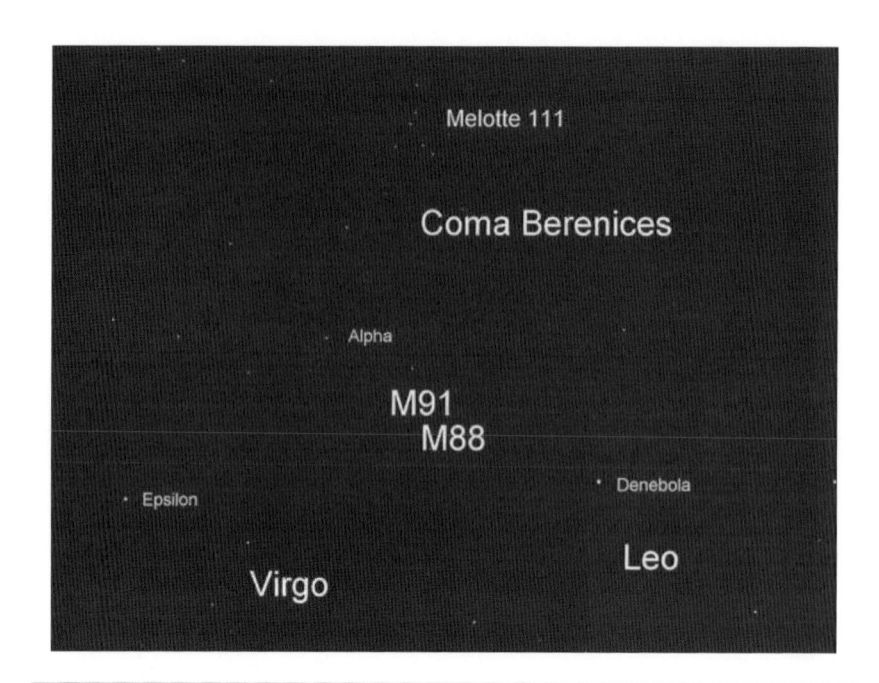

Fig. 6.53 How to find M88

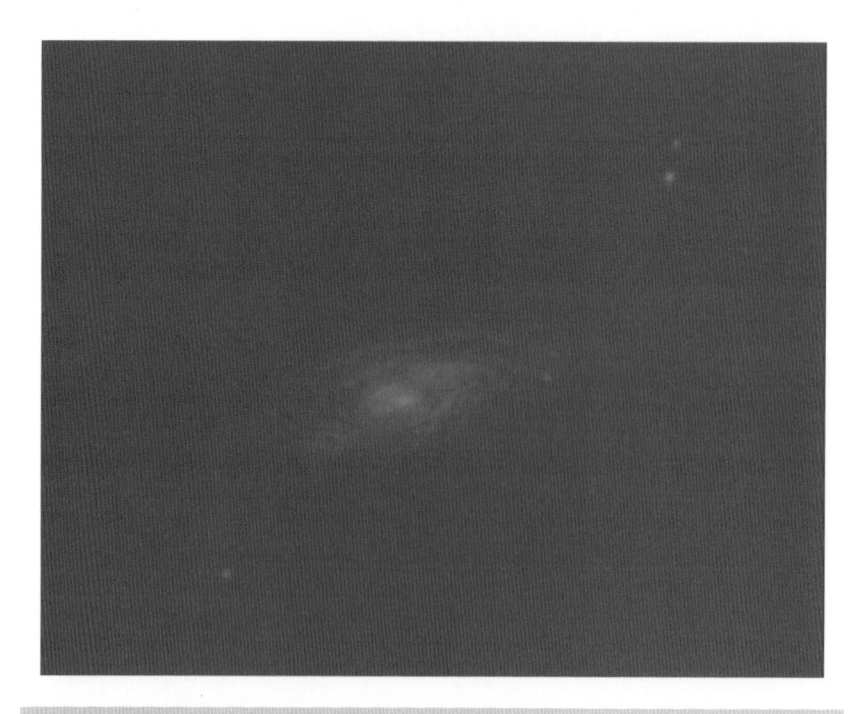

Fig. 6.54 M88 modified (NASA/courtesy of nasaimages.org)

M89

M89 is an elliptical galaxy in Virgo and a member of the Virgo Supercluster. It was discovered by Charles Messier in 1781 and is 4 arcminutes across, suggesting a diameter of about 60 000 light years at its distance of 60 million light years. It appears to be immersed in an envelope of intergalactic gas.

Where to Find

M89 is one of the faintest Messier objects and is best found using M87 as a guidepost, then moving east from M84 and M87. It is not visible in a finderscope (Fig. 6.55).

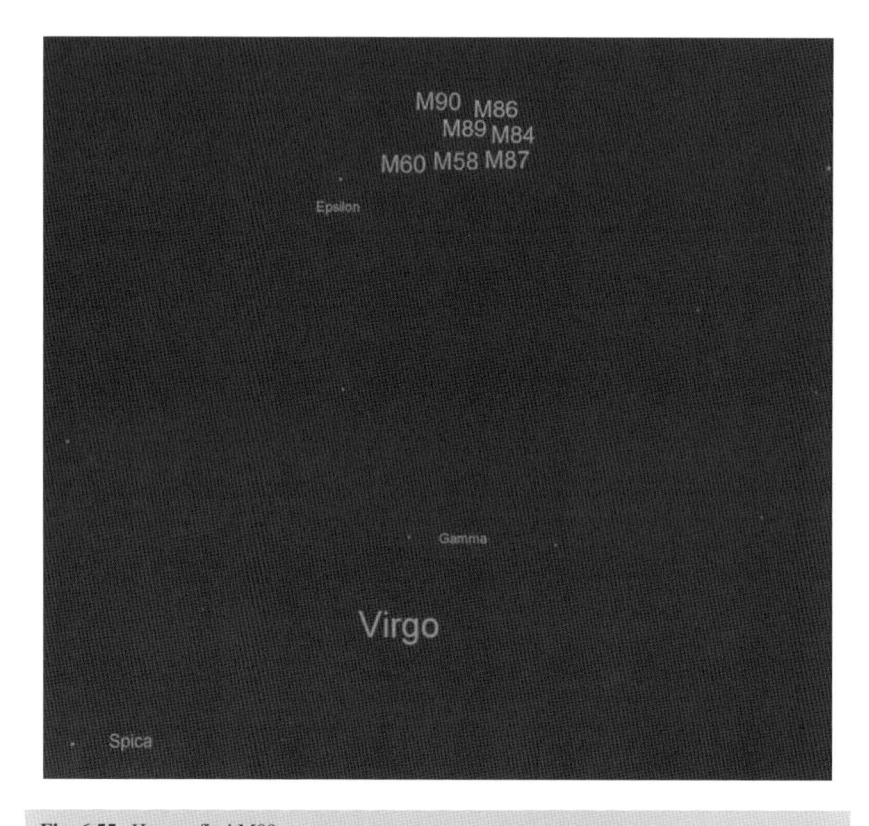

Fig. 6.55 How to find M89

Object Type	Elliptical galaxy
Declination	12° 33 min
Right Ascension	12 h 35.7 min
Magnitude	9.8

What It Looks Like

M89 appears as a small fuzzy patch to a 127 mm Maksutov in good conditions and is on the borderline of resolution, showing no obvious structure.

Charles Messier's Original Notes

18 March 1781
(RA: 12 h 24 m 38 s, 186d 09′ 36″; Dec: +13d 46′ 49″; Diameter: –)

Nebula without star, in Virgo, a short distance from & on the same parallel as the nebula reported above, No. 87 [M87]. Its light is extremely faint & rare, & it is not without struggle that we can see [it].

How to View

M89 appears unspectacular with a modest instrument and really needs a large aperture telescope to do it justice.

Photographic Details

Figures 6.56 and 6.57

Fig. 6.56 M89 (NASA/courtesy of nasaimages.org)

Fig. 6.57 M89 (ModifiedNASA/courtesy of nasaimages.org)

M90

M90 is a spiral galaxy in Virgo. It was discovered in 1781 by Charles Messier. It is a member of the Virgo Supercluster, sharing a common distance of 60 million light years and it has a large angular size of 9.5×4.5 arcminutes. However, there has been some doubt as to its membership of the supercluster, as it is moving very quickly towards us. Strangely for a spiral galaxy, star formation appears to have stopped in its spiral arms.

Where to Find

M90 is a tough object but is just under a degree from M89 and almost due north, so you should find M89 first, via M87 and M84/M86, then move north until M90 comes into view. M90 is slightly brighter than M89 and may be glimpsed in the same field of view if you can achieve a field of view of at least 90 arcminutes (Fig. 6.58).

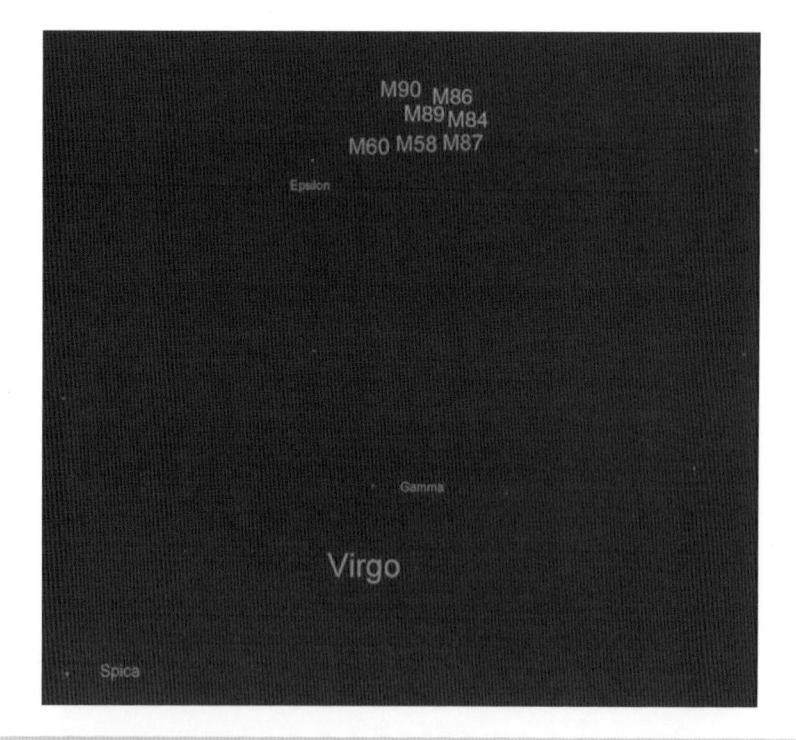

Fig. 6.58 How to find M90

M90

Object Type	Spiral galaxy
Declination	13° 10 min
Right Ascension	12 h 36.8 min
Magnitude	9.5

What It Looks Like

Although M90 is a spiral galaxy, a 127 mm Maksutov only reveals it as a fuzzy patch of indeterminate nature in the neighbourhood of M87.

Charles Messier's Original Notes

18 March 1781
(RA: 12 h 25 m 48 s, 186d 27' 00"; Dec: +14d 22' 50"; Diameter: –)

Nebula without star, in Virgo: its light is as faint as the previous, No. 89 [M89].

How to View

M90 appears unspectacular with a modest instrument and really needs a large aperture telescope to do it justice.

Photographic Details

Figures 6.59 and 6.60

Fig. 6.59 M90 (NASA/courtesy of nasaimages.org)

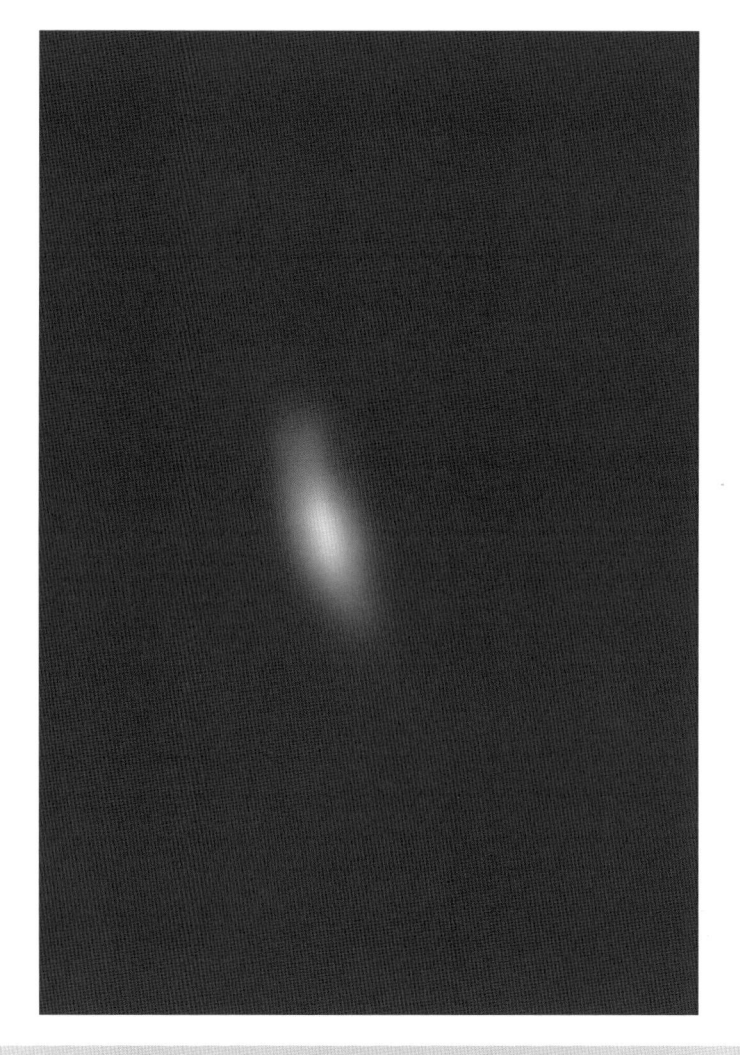

Fig. 6.60 M90 modified (NASA/courtesy of nasaimages.org)

M91

M91 is a spiral galaxy in Coma Berenices. It was discovered by Charles Messier in 1781 and lost again, as he failed to record its position accurately. It was rediscovered by William Herschel in 1784. It is a member of the Virgo Supercluster and shares its common distance of 60 million light years. It is 5.4×4.4 arcminutes across but is actually the faintest member of the Messier Catalogue.

Where to Find

Please refer to Fig. 6.52, as M91 is close to M88.

Object Type	Spiral galaxy
Declination	14° 30 min
Right Ascension	12 h 35.4 min
Magnitude	10.2

What It Looks Like

Although M91 is a spiral galaxy, it appears as a small fuzzy patch of indeterminate nature in a 127 mm Maksutov on a good night. It forms a close pair with M88.

Charles Messier's Original Notes

18 March 1781
(RA: 12 h 26 m 28 s, 186d 37' 00"; Dec: +14d 57' 06"; Diameter: –)

Nebula without star, in the Virgin [Virgo], above the preceding one, No. 90 [M90]: its light is even fainter than that of the above.

Note. The constellation of Virgo, & above all, the northern wing, is one of the constellations that encloses the most nebulae: This catalogue contains 13 that have been determined: namely, No's. *49, 58, 59, 60, 61 , 84, 85, 86, 87, 88, 89, 90 & 91.* All these nebulae appear to have no stars: you cannot see them unless there is a very good sky, & [they are] near their Meridian passages. Most of these nebulae had been given to me by M. Méchain.

How to View

M91 appears unspectacular with a modest instrument and really needs a large aperture telescope to do it justice.

Photographic Details

Figures 6.61 and 6.62

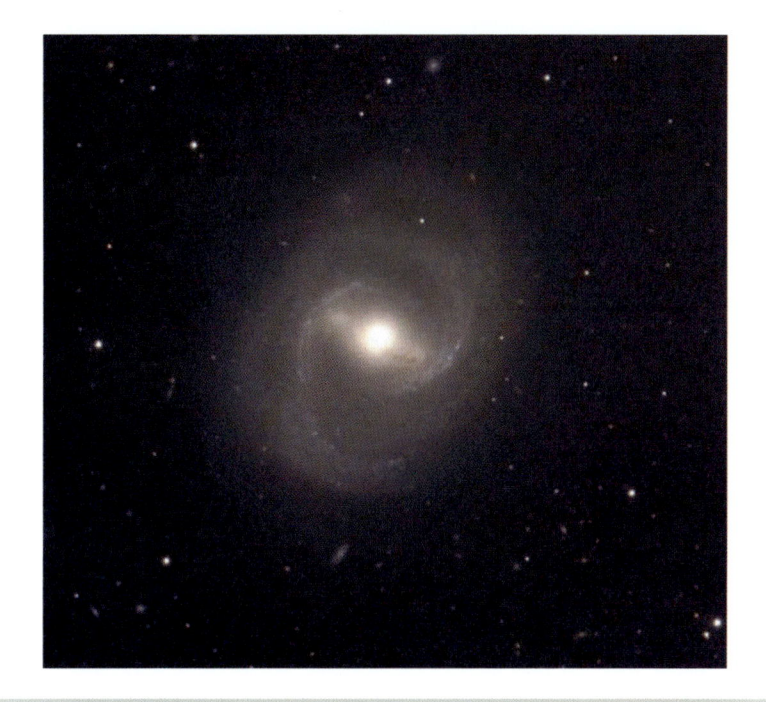

Fig. 6.61 M91 (NASA/courtesy of nasaimages.org)

Fig. 6.62 M91 modified (NASA/courtesy of nasaimages.org)

Chapter 7

M92–M110

M92

M92 is a globular star cluster in Hercules and it is in the northern part. It is often overlooked in favour of its more famous neighbour M13 but is half as bright and better placed from the northern hemisphere than M13, being circumpolar from northern Europe and most of Canada. It was discovered in 1777 by Johann Elert Bode and is about 14 arcminutes across and between 26,000 and 27,000 light years away.

Although I'd seen M92 before, I was browsing at the double stars in the Draco region when I saw a bright object resembling a comet without a tail. Remembering the original reason for the Messier Catalogue, I checked my star atlas and confirmed that it was M92 but it was much brighter than usual, due to almost perfect seeing conditions. Thank you, Charles Messier, for saving me from the embarrassment that you saved yourself.

Where to Find

M92 is in a rather faint part of sky and there are plenty of faint background stars to cause confusion. Fortunately, a good finderscope can pick it up. One way to find the general area is to find M13 then move about 8° in the direction of Gamma Draconis (Fig. 7.2).

Fig. 7.1 M92 by Anthony Glover

Object Type	Globular star cluster
Declination	43° 08 min
Right Ascension	17 h 17.1 min
Magnitude	6.4

What It Looks Like

M92 can be rather enigmatic. Most nights it appears as a rather unremarkable fuzzy patch in 70 mm binoculars and a 127 mm Maksutov does little to improve the view. If you can catch it on an exceptionally clear night, it is more interesting but not so spectacular as the nearby M13. The faint outer regions and denser core are obvious, though.

Charles Messier's Original Notes

18 March 1781
(RA: 17 h 10 m 32 s, 257d 38′ 03″; Dec: +43d 21′ 59″; Diameter: 0d 05′)

Nebula, beautiful, distinct, & very bright, between the knee & the left leg of Hercules, in very good in a telescope of one foot. It contains no star; the center is clear & brilliant, surrounded by nebulosity & resembles the nucleus of a large comet: its light, its size, approaches much of that of the nebula that is in the belt of Hercules. See No. 13 of this catalogue: its position was determined by comparing it directly to the star σ Herculis, 4th magnitude: the nebula & the star lie along the same parallel.

How to View

With a small instrument, M92 shows some structure. The view is often improved by use of a light pollution reduction filter. To resolve some outer cluster members, you need a larger aperture telescope.

Photographic Details

M92 was taken using a Skywatcher ED100, semi-APO, SXV H9, with an infrared blocking filter. It consisted of 15 1 min exposures (Figs. 7.1 and 7.3).

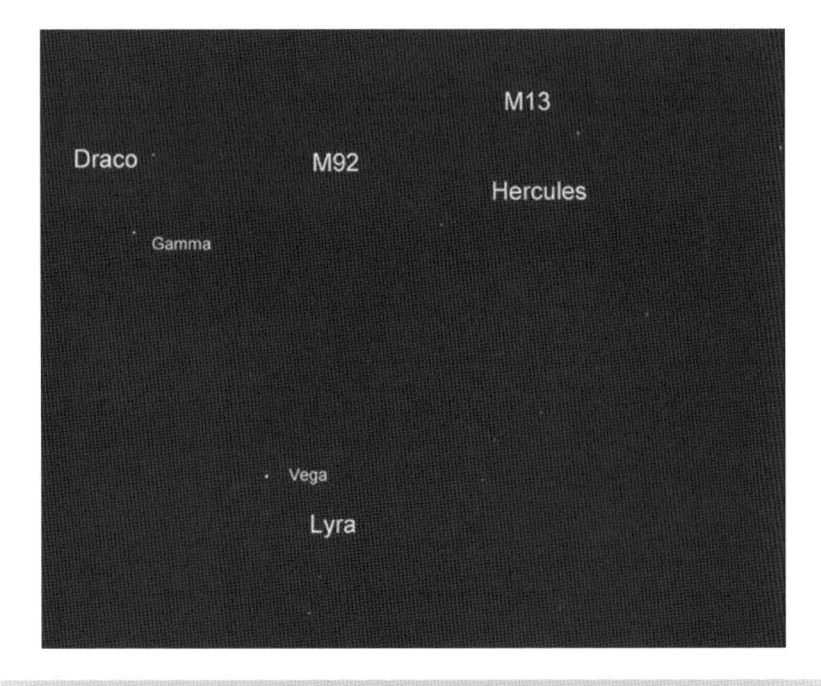

Fig. 7.2 How to find M92

Fig. 7.3 M92 modified

M93

M93 is an open star cluster in Puppis. It was discovered by Charles Messier in 1781. It is 22 arcminutes across and 3,600 light years away. It has about 80 members.

Where to Find

M93 is found in the southern part of Puppis near the main asterism. Fortunately, it is quite bright at magnitude six, so can be seen on a reasonable night from suburban skies. Puppis itself may not always be visible to the unaided eye and it is to the east of Canis Major. The nearest bright star is Xi Puppis and M93 is a degree or so to the north west.

Object Type	Open star cluster
Declination	−23° 52 min
Right Ascension	7 h 44.6 min
Magnitude	6

What It Looks Like

M93 is a difficult object from northern Europe and Canada, as it never gets particularly high. It is a different prospect south of the equator and is quite unmistakable even in small binoculars. However even large binoculars show it as a faint blur and a 127 mm Maksutov just starts to resolve some of the brighter (10th magnitude) stars, which blink in and out of view with scintillation. The brighter stars form a roughly triangular shape.

Charles Messier's Original Notes

20 March 1781
(RA: 07 h 35 m 14 s, 113d 48′ 35″; Dec: −23d 19′ 45″; Diameter: 0d 08′)

Cluster of small stars, without nebulosity between the Great Dog [Canis Major] & the bow of the ship [Puppis].

Fig. 7.4 How to find M93

Fig. 7.5 M93 (NASA/courtesy of nasaimages.org)

Fig. 7.6 M93 modified (NASA/courtesy of nasaimages.org)

How to View

Firstly, if you are viewing from the northern hemisphere, you need to wait until it is due south, as extinction is a huge problem. There is no problem with using large apertures and longer focal lengths because you only need a field of view of about 30 arcminutes to see its full extent clearly.

Photographic Details

Figures 7.5 and 7.6

M94

M94 is a spiral galaxy in Canes Venatici. It was discovered by Pierre Méchain in 1781. It is 14.5 million light years away and measures 7×3 arcminutes. It has two rings of star formation suggesting a recent close encounter with another galaxy or a merger.

Where to Find

M94 lies near the two brightest stars of Canes Venatici. It is about 2° north east of a line joining them. However, with a faint magnitude of 8.2, it is unlikely to be seen in a finderscope, except under excellent conditions from a dark site (Fig. 7.8).

Object Type	Spiral galaxy
Declination	41° 07 min
Right Ascension	12 h 50.9 min
Magnitude	8.2

What It Looks Like

M94 appears as a fuzzy patch to 70 mm binoculars on a good night. Some hint of the spiral structure is visible but it was fuzzy and the spiral arms kept blinking in and out of view when using the light pollution reduction filter with a 127 mm Maksutov. Without the filter, no structure is discernable.

Charles Messier's Original Notes

24 March 1781

Fig. 7.7 M94 by Anthony Glover

(RA: 12 h 40 m 43 s, 190d 10′ 46″; Dec: +42d 18′ 43″; Diameter: 0d 02.5′)
M. Méchain: (RA: 190d 09′ 38″; Dec: +42d 18′ 50″)

Nebula without star, above the heart of Charles [α Canum Venaticorum], on the same parallel as the star No. 8 [β Canum Venaticorum], of 6th magnitude of the Hunting Hounds [Canes Venatici], Flamsteed follows that: the centre is brilliant & has somewhat diffuse nebulosity. It resembles the nebula which is below the Hare [Lepus], No. 79 [M79], but it is brighter & more beautiful: M. Méchain made the discovery of this on 22 March 1781.

How to View

M94 appears much easier than galaxies of the Virgo Supercluster from the northern hemisphere, as it can be just about directly overhead. The light pollution reduction filter was a definite help but more aperture would be even better.

Photographic Details

This photograph of M94 was taken using an SBIG ST2000XCM cooled to −25° with a 80 mm APO Refractor and Astronomik LPR filter. 10 5 min subframes were stacked (Fig. 7.7).

Figure 7.9 shows how M94 appears through a 127 mm Maksutov.

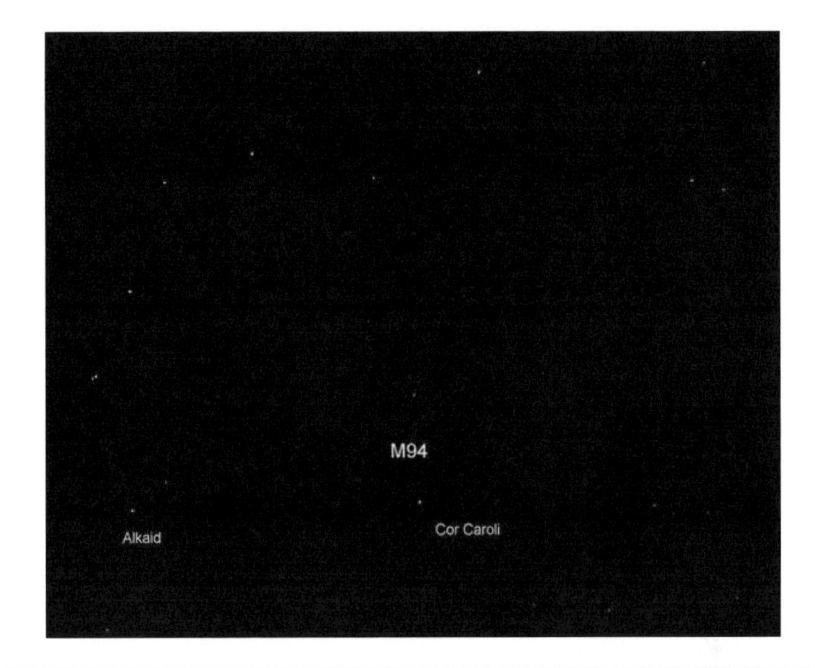

Fig. 7.8 How to find M94

Fig. 7.9 M94 modified

M95

M95 is a spiral galaxy in Leo. It was discovered by Pierre Méchain in 1781. It is close to M96, M105 and several smaller galaxies, forming the Leo I Group. It is 30 million light years away and measures 4.4×3.3 arcminutes across. It is a barred spiral galaxy and its spiral arms are nearly circular.

Where to Find

Unfortunately, M95 is one of the toughest objects in the Messier Catalogue to find. Not only is it a faint magnitude 9.7 spiral galaxy but there is no nearby bright object. It might be easier to find M96 first and then move about a degree to the west. It is unlikely to be visible in many finderscopes (Fig. 7.10).

Object Type	Spiral galaxy
Declination	11° 42 min
Right Ascension	10 h 44.0 min
Magnitude	9.7

What It Looks Like

M95 is more of an object to cross off the list than to be enjoyed. It appears as a small fuzzy patch and it is not possible to ascertain what type of object it is through a 127 mm Maksutov and Méchain deserves a lot of credit for having found it at all.

Fig. 7.10 How to find M95

Charles Messier's Original Notes

24 March 1781
(RA: 10 h 32 m 12 s, 158d 03' 05"; Dec: +12d 50' 21"; Diameter: –)
M. Méchain: (RA: 158d 06' 23"; Dec: +12d 49' 50")

Nebula without star, in the Lion [Leo], above the star 1 [Leonis]: its light is very faint.

How to View

Through modest instruments, you need a clear night and Leo must be near the meridian to stand any chance of seeing it at all. If you own or otherwise have access to a larger aperture telescope, it is best to use it and a dark site helps as well.

Photographic Details

Figures 7.11 and 7.12

Fig. 7.11 M95 (NASA/courtesy of nasaimages.org)

Fig. 7.12 M95 modified (NASA/courtesy of nasaimages.org)

M96

M96 is a spiral galaxy in Leo and is a member of the Leo I Group, along with M95, M105 and other smaller galaxies. It is 38 million light years away and measures 6×4 arcminutes across. It was discovered by Pierre Méchain in 1781. Its apparent size suggests diameter of 66,000 light years but faint extensions suggest it is 100,000 light years across, about the same as our Milky Way.

Where to Find

M96 is a hard object to find but slightly easier than the nearby M95, being half a magnitude brighter. It is just under 2° and very slightly west of the 6th magnitude

M96

Fig. 7.13 M96 (NASA/courtesy of nasaimages.org)

star 53 Leonis. I would suggest that if you cannot see M65 and M66 clearly, it is not worth trying for either M95 nor M96 (Fig. 7.14).

Object Type	Spiral galaxy
Declination	11° 49 min
Right Ascension	10 h 46.8 min
Magnitude	9.2

What It Looks Like

M96 is near the limit of what can be seen with my set-up. Although it is a tough object that you need to stare at for a few minutes, it is definitely different to its neighbour, M95. At least to my eyes and equipment, its overall impression reminded me of the Pinwheel (M33), although much fainter.

Charles Messier's Original Notes

24 March 1781
(RA: 10 h 35 m 05 s, 158d 46' 20"; Dec: +12d 58' 09"; Diameter: –)
M. Méchain: (RA: 158d 48' 00"; Dec: +12d 57' 33")

Nebula without star, in the Lion [Leo], near the previous [M95]: though it is less distinct, both are on the same parallel of *Regulus:* they resemble the two nebulae of the Virgin [Virgo], No's. 84 & 86 [M84 and M86]. M. Méchain saw them both on 20 March 1781.

How to View

Through modest instruments, you need a clear night and Leo must be near the meridian to stand any chance of seeing it at all. If you own or otherwise have access to a larger aperture telescope, it is best to use it and a dark site helps as well.

Photographic Details

Figures 7.13 and 7.15

Fig. 7.14 How to find M96

Fig. 7.15 M96 modified (NASA/courtesy of nasaimages.org)

Fig. 7.16 M97 by Anthony Glover

M97

M97 is known as the Owl Nebula and is a planetary nebula in Ursa Major. It is 4.3×3.3 arcminutes across and is about 2,600 light years away, as far as these distances can be determined but there's estimates from about half that distance to over 10,000 light years. It was discovered by Pierre Méchain in 1781. The central star is about 16th magnitude and believed to be about 0.7 solar masses. M97 is circumpolar from much of northern Europe and Canada.

Where to Find

M97 is quite close to Beta Ursae Majoris, about 2° along a line joining Beta and Gamma and very slightly south. In fact, it is in the same binocular or finderscope field of view as Beta, although it is unlikely to be seen, as it is quite faint.

Object Type	Planetary nebula
Declination	55° 1 min
Right Ascension	11 h 14.8 min
Magnitude	9.9

What It Looks Like

M97 can be seen as a misty patch in a 127 mm Maksutov. It is too faint to determine the type of object but there's a suggestion that the shape is oval. It is likely to be invisible, except on a clear night from a dark site or with a light pollution reduction filter.

Charles Messier's Original Notes

24 March 1781
(RA: 11 h 01 m 15 s, 165d 18' 40"; Dec: −56d 13' 30" [sic]; Diameter: 0d 02')

Nebula in Ursa Major, near β [Ursa Majoris]: it is difficult to see, reports M. Méchain, especially when the micrometer is illuminated: its light is faint and without a star. M. Méchain saw [it] for the first time on 16th February 1781, & the position is reported by him. Near this nebula there is another which has not yet been determined [positionally], and a third which is close to γ of the Great Bear [Ursa Majoris].

How to View

As M97 is one of the fainter Messier objects, even a favourable declination (from the northern hemisphere) and light pollution are little substitutes for aperture.

M97   317

Fig. 7.17 How to find M97

Fig. 7.18 M97 modified

Photographic Details

M97 was taken using seven 10 min exposures with an 110 mm/F7 refractor, SBIG St2000×cm cooled to −25°, dark subtracted, stacked and processed in Nebulosity. M108 is at the bottom right (Fig. 7.16).

M97 is on the verge of visibility with a 127 mm Maksutov, being the faintest of the Messier objects (Fig. 7.18).

M98

M98 is a spiral galaxy in Coma Berenices. It was discovered by Pierre Méchain in 1781. It is part of the Virgo Supercluster and shares its common distance of 60 million light years. It is 9.5×3.2 arcminutes across, suggesting a large size but is quite faint as a lot of dust lies in our line of sight. It forms a group with M99 and M100.

Where to Find

Although, strictly speaking, M98 and its neighbours M99 and M100 lie in the faint constellation of Coma Berenices, the ideal starting point for a search is in the neighboring constellation of Leo. The background star Delta Comae Berenici is about 4° east of Denebola (Gamma Leonis) and M98 is just to the west of Delta. However, as M98 is one of the fainter Messier objects, it is not visible in a finderscope and even M100 (the

Fig. 7.19 M98 (NASA/courtesy of nasaimages.org)

M98

Object Type	Spiral galaxy
Declination	14° 54 min
Right Ascension	12 h 13.8 min
Magnitude	10.1

brightest of the trio at magnitude 9.3) is hard to spot. It is worth doing a quick test on the nearby M65 and M66 because if you cannot see them with your equipment under the prevailing conditions, you will not see M98 nor its neighbors (Fig. 7.20).

What It Looks Like

M98 is one of the fainter Messier objects. Although it is detectable in a 127 mm Maksutov in ideal conditions, its spiral structure is far from obvious, yet it has a bright nucleus. It is fainter than the nearby M99 and M100.

Charles Messier's Original Notes

13 April 1781
(RA: 12 h 03 m 23 s, 180d 50' 49"; Dec: +16d 08' 15"; Diameter: –)

Nebula without star, an extremely faint light, above the northern wing of Virgo, on the same parallel & close to the star No. 6 [6 Comae Berenices], of 5th magnitude, of the hair of Berenice [Coma Berenices], according to Flamsteed. M. Méchain saw it on 15th March 1781.

How to View

As M98 is one of the fainter Messier objects, even a favourable declination (from the northern hemisphere) and light pollution are little substitutes for aperture. To see the spiral structure properly, the suggested minimum aperture is 200 mm.

Photographic Details

Figures 7.19 and 7.21

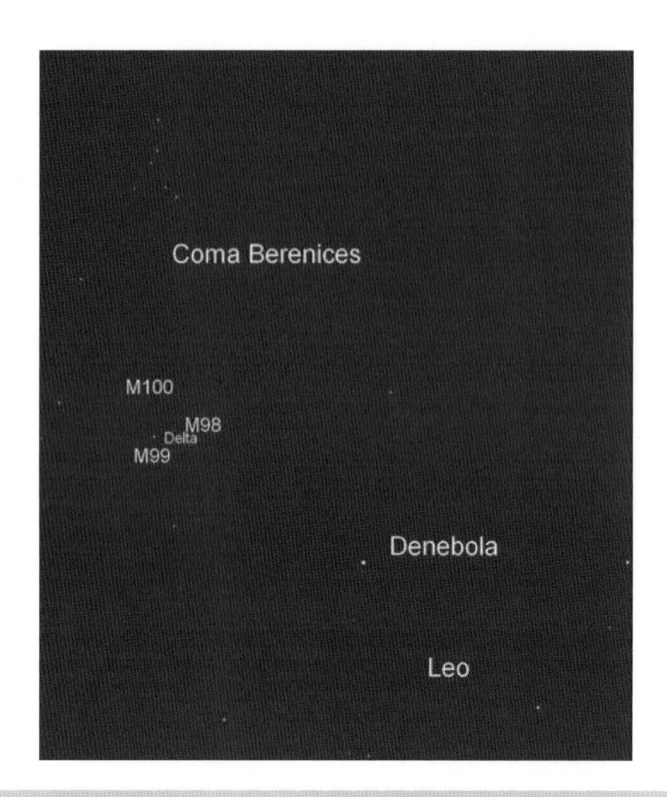

Fig. 7.20 How to find M98

Fig. 7.21 M98 modified (NASA/courtesy of nasaimages.org)

M99

M99 is a spiral galaxy in Coma Berenices. It was discovered by Pierre Méchain in 1781. It is part of the Virgo Supercluster and shares its common distance of 60 million light years. It is 5.4×4.8 arcminutes across. It forms a group with M98 and M100. It has probably interacted with at least one of its neighbors, as its nucleus is somewhat off-center.

Where to Find

Please refer to Fig. 7.20, as M99 is in the same area.

Object Type	Spiral galaxy
Declination	14° 25 min
Right Ascension	12 h 18.8 min
Magnitude	9.9

What It Looks Like

M99 is one of the fainter members of the Messier Catalogue, yet its extra 0.2 magnitude of brightness makes it more readily visible than the nearby M98. The view through a 127 mm Maksutov in good conditions suggests that something "fuzzy" is there but does not conclude whether it is a spiral or elliptical galaxy.

Charles Messier's Original Notes

13 April 1781
(RA: 12 h 07 m 41 s, 181d 55' 19"; Dec: +15d 37' 12"; Diameter: –)

Nebula without star, with a very faint, rare light, but a little clearer than the previous one, placed along the northern wing of the Virgin [Virgo], & near the same star, No. 6, of Coma Berenices. The nebula is between two stars of 7th & 8th magnitude. M. Méchain saw it on 15th March 1781.

How to View

As M99 is one of the fainter Messier objects, even a favourable declination (from the northern hemisphere) and light pollution reduction filter are little substitutes for aperture. To see the spiral structure properly, the suggested minimum aperture is 180 mm.

Fig. 7.22 M99 (NASA/courtesy of nasaimages.org)

Photographic Details

Figures 7.22 and 7.23

M100

M100 is a spiral galaxy in Coma Berenices. It was discovered by Pierre Méchain in 1781. It is part of the Virgo Supercluster and shares its common distance of 60 million light years. It is 7×6 arcminutes across, suggesting a large size. It forms a group with M98 and M99. It is undergoing a burst of star formation, possibly due to gravitationally interaction with its neighbors.

M100

Fig. 7.23 M99 modified (NASA/courtesy of nasaimages.org)

Where to Find

Please refer to Fig. 7.20, as M100 is in the same area.

Object Type	Spiral galaxy
Declination	15° 49 min
Right Ascension	12 h 22.9 min
Magnitude	9.3

What It Looks Like

Although M100 isn't one of the easier Messier objects, it is noticeably brighter than its neighbors, M98 and M99 and there is a definite suggestion of a spiral structure with a bright nucleus.

Charles Messier's Original Notes

13 April 1781
(RA: 12 h 11 m 57 s, 182d 59' 19"; Dec: +16d 59' 21"; Diameter: –)

Nebula without star, with the same light as the previous one, found in the ear of the Virgin [Virgo]. Viewed by Méchain on 15th March 1781. These three nebulae, No's. 98, 99 & 100, are very difficult to recognize because of their faintness: we can only see them in fine weather, & close to the passage of their Meridian.

How to View

Fortunately, M100 is slightly less demanding than its neighbors, so a 127 mm Maksutov, with suitably low magnification and a light pollution reduction filter will do the job – just!

Photographic Details

Figure 7.24 shows M100 and some faint companions in the Virgo Supercluster. It was taken using a 80 mm/F6 refractor and SBIG ST2000×cm camera cooled to –25°. Eight images, each of 5 min duration were taken and stacked and processed using *Nebulosity* (Stark Software).

The view through a 127 mm Maksutov (as shown in Fig. 7.25) isn't as clear but the spiral arms can be seen.

Fig. 7.24 M100 by Anthony Glover

Fig. 7.25 M100 modified

M101

M101 is a spiral galaxy in Ursa Major. It was discovered by Pierre Méchain in 1781. It is 22 arcminutes across and 27 million light years distant. This gives a true diameter of 170,000 light years, much larger than our own Milky Way. It is surrounded by some satellite galaxies but none of them is in the Messier Catalogue. M101 is circumpolar from much of Europe and Canada but is best seen in spring.

Where to Find

M101 forms an approximate right angled triangle with Eta and Zeta Ursae Majoris (Mizar). It is east of Mizar and north of Eta Ursae Majoris by about 5°. It can be quite hard to find, as most nights it is too faint to see in a finderscope. A wide field of view in the main telescope is almost essential.

Object Type	Spiral galaxy
Declination	54° 21 min
Right Ascension	14 h 3.2 min
Magnitude	7.9

What It Looks Like

M101 is quite bright but its brightness is spread out over a large area. It isn't visible in 70 mm binoculars but it can be seen in a 127 mm Maksutov-Cassegrain. Its spiral structure is far from obvious, though, and it appears more like a central condensation surrounded by some wispy nebula, so would suggest an elliptical galaxy instead.

The addition of a light pollution reduction filter transforms the view into that of a classic spiral galaxy and it was amazing to see just how large it can appear when its spiral arms are visible.

Pierre Méchain's Original Notes

27 March 1781
(RA: 13 h 43 m 28 s, 208d 52' 04"; Dec: +55d 24' 25"; Diameter: 0d 07')

Nebula without star, very obscure & very wide, six to seven [arc]minutes in diameter, between the left hand of Boötes & the tail of the Great Bear [Ursa Major]. It is difficult to distinguish when illuminating the thread [of the micrometer].

M101

Fig. 7.26 How to find M101

How to View

M101 does not respond well to high magnification, as it spreads the light of the other regions into too much space. It is best to use a telescope/eyepiece combination that yields a field of view of at least 1°. It is unlikely to be visible to binoculars or small aperture telescopes, except under very good conditions. With a small/medium aperture, a light pollution reduction filter or clear sky is necessary to see the spiral arms. Of the galaxies in the Messier Catalogue, M101 is one of the more realistic targets for a 125 mm aperture.

Photographic Details

Figure 7.27 was taken using a 10″ F5 Newtonian, Starlight Xpress H×916, guided with a 80ED refractor with a modifed Toucam. It consisted of 29 5 min exposure luminance frames, ten 3 min exposures of hydrogen alpha frames (binned 2×2) and 5 5-min exposures of RGB.

Figure 7.28 shows M101 as it appears through my Maksutov but with the light pollution reduction filter. Without it, only the core is visible.

Fig. 7.27 M101 by Mike Deegan

Fig. 7.28 M101 modified

M102

There is some confusion as to whether M102 is a duplicate entry of M101 or is a separate entry. Notes by Charles Messier seem to indicate that it is the lenticular galaxy (NGC5866) and it was found either by Charles Messier or Pierre Méchain. Certainly NGC 5866 isn't the faintest object in the Messier Catalogue but then there's several objects that are more observable from Paris skies than some of the actual entries.

It is about 45 million light years away in Draco and has dimensions of 5.2×2.3 arcminutes. It has a conspicuous dust lane and is about 69,000 light years across.

Where to Find

M102 is a tough object to find. There are no obvious guideposts, with Iota Draconi being over 4° away. It is not visible in finderscopes (Fig. 7.30).

Fig. 7.29 M102 (NASA/courtesy of nasaimages.org)

Object Type	Lenticular galaxy
Declination	55° 46 min
Right Ascension	15 h 6.5 min
Magnitude	9.9

What It Looks Like

M102 is invisible in 70 mm binoculars and only appears as an out of focus "star" in a 127 mm Maksutov.

Pierre Méchain's Original Notes

[No coordinates or date in published catalogue]

Nebula near the stars o Boötis and ι Draconis: it is very faint, it is near a star of 6th magnitude. [Méchain later came to the view that M102 was in fact a mistaken re-observation of M101. In later years however, many astronomers wondered whether M102 was a distinct object after all. The common (though not yet official) consensus now is that M102 is indeed distinct, and is in fact the Spindle Galaxy NGC5866].

How to View

As this is one of the fainter Messier objects, it really needs a large aperture to see properly but can be detected in a modest telescope.

Photographic Details

Figures 7.29 and 7.31

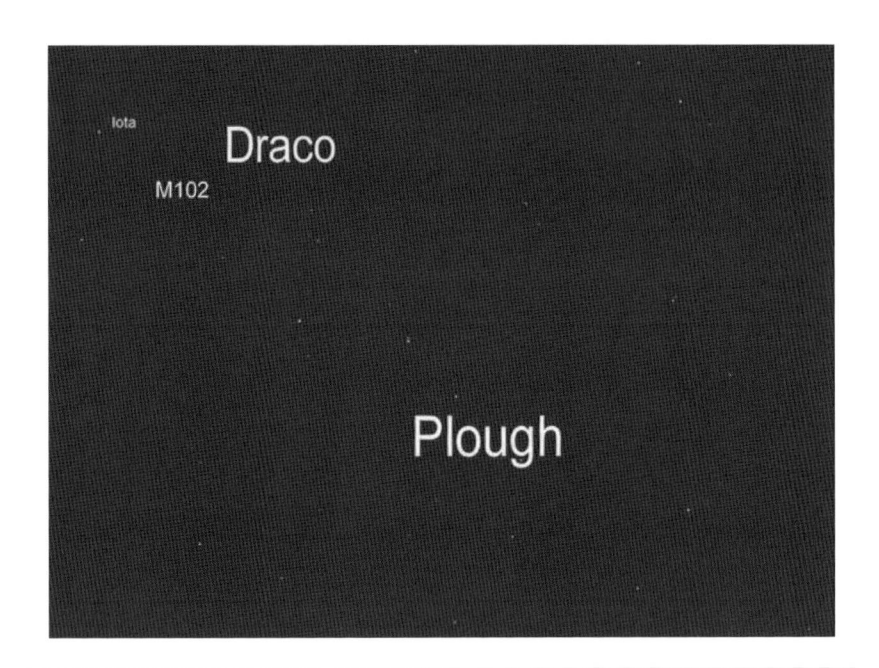

Fig. 7.30 How to find M102

Fig. 7.31 M102 modified (NASA/courtesy of nasaimages.org)

M103

M103 is an open star cluster in Cassiopeia. It is 8,500 light years away and contains anything from 25 to 173 member stars, depending on who you believe. Indeed, its existence was challenged as just a line of sight effect along the Milky Way but it has been proved that at least some stars in the field of 6 arcminutes share a common motion. It can be seen as a fuzzy patch in binoculars and is circumpolar from much of Europe and the USA.

Where to Find

M103 is about 1° north east of Delta Cassiopei. It is quite small, so sometimes difficult to pick up in finderscopes but if you use a low power eyepiece, you can get it in the same field of view of the star (Fig. 7.33).

Object Type	Open star cluster
Declination	60° 42 min
Right Ascension	1 h 33.2 min
Magnitude	7.4

What It Looks Like

M103 appears as a group of brighter stars against the Milky Way background. As it is fainter and smaller than the nearby M52, it is quite a challenge in binoculars

Fig. 7.32 M103 by Anthony Glover

against the background stars. On a good night, my finderscope picks it up and the stars are prominent against the Milky Way background stars through my Maksutov. There seems no obvious shape to this cluster.

Pierre Méchain's Original Notes

[No coordinates or date in published catalogue]
  Star cluster between ε & δ of the leg of Cassiopeia.

How to view

As M103 is not particularly bright, it benefits from "aperture fever" and the bigger the aperture the better. As its is 6 arcminutes across, it can take quite a bit of magnification, around 200x, more with a wide field of view eyepiece.

Photographic Details

This is also a good representation of how it appears in a 127 mm Maksutov (Fig. 7.32).

Fig. 7.33 How to find M103

M104

M104 is a spiral galaxy in Virgo. It was discovered by Pierre Méchain in 1781. It is known as the Sombero Galaxy, as it is almost edge on and its dark dust lane gives the appearance of a hat. It is 9×4 arcminutes in size and about 50 million light years away and not a member of the Virgo Supercluster. It is the dominant member of a small cluster of its own. It has a pronounced bulge and has an unusually high number of globular star clusters for a spiral galaxy.

Where to Find

M104 is in a sparse patch of sky and its low declination makes it very difficult to find. It forms a rough right angled triangle with Spica (Alpha Virginis) and Porrima (Gamma Virginis) (Fig. 7.35).

Object Type	Spiral galaxy
Declination	−11° 37 min
Right Ascension	12 h 40.0 min
Magnitude	8

Fig. 7.34 M104 (Courtesy of Ray Grover, Manchester Astronomical Society)

What It Looks Like

M104 is invisible in 70 mm binoculars from my home (normal viewing location) but shows a hint of structure with the Maksutov but not enough to give the classic look shown in photographs.

How to View

M104 isn't one of the fainter objects but its southerly declination makes it a difficult object from mid-northern altitudes. A light pollution reduction filter helps but apart from that, aperture is the most significant factor.

Photographic Details

Figure 7.34 is a good representation of how M104 appears in a 127 mm Maksutov.

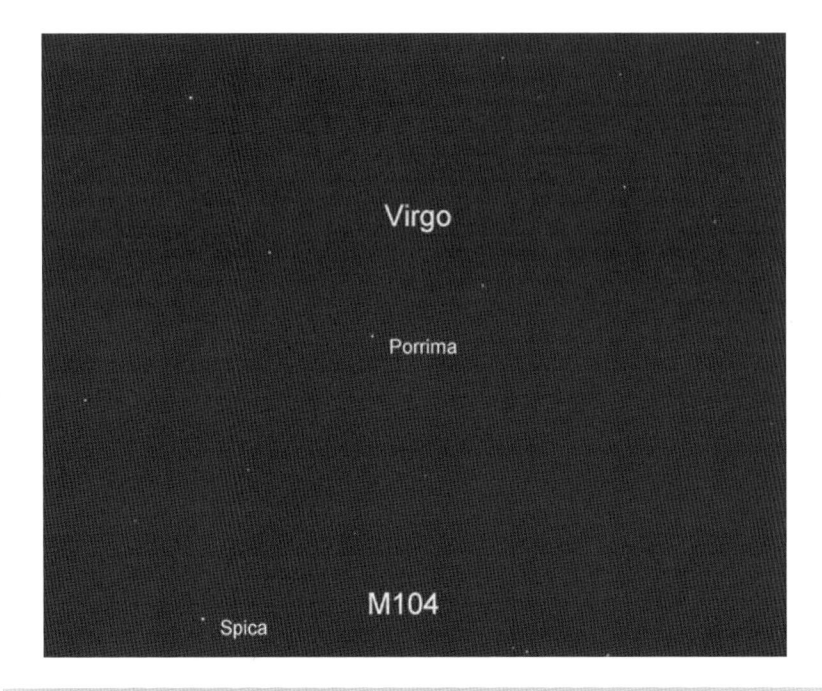

Fig. 7.35 How to find M104

M105

M105 is an elliptical galaxy in Leo. It is a member of the Leo I Group and shares its common distance of about 38 million light years. It is 2 arcminutes across and was discovered by Pierre Méchain in 1781. It has a central massive object of about 20 million solar masses, probably a black hole.

Where to Find

M105 is quite difficult to find and is about 2° north of the 6th magnitude star 53 Leonis. As it is a very similar brightness to M96 and about a degree away, it is possible to mistake one for the other (Fig. 7.37).

Fig. 7.36 M105 (NASA/courtesy of nasaimages.org)

M105

Object Type	Elliptical galaxy
Declination	12° 35 min
Right Ascension	10 h 47.8 min
Magnitude	9.3

What It Looks Like

M105 appears as a faint object and not dissimilar to a globular star cluster in a 127 mm Maksutov. Its apparent size suggests that only its central parts near the nucleus are visible.

How to View

Through modest instruments, you need a clear night and Leo must be near the meridian to stand any chance of seeing it at all. If you own or otherwise have access to a larger aperture telescope, it is best to use it and a dark site helps as well.

Photographic Details

Figures 7.36 and 7.38

Fig. 7.37 How to find M105

Fig. 7.38 M105 modified (NASA/courtesy of nasaimages.org)

M106

M106 is a spiral galaxy in Canes Venatici. It was discovered by Pierre Méchain in 1781. It is 19×8 arcminutes across and 25 million light years away. It is not really known whether it is a member of any galaxy groups or is on its own. There is a high incidence of hot, blue stars, suggesting recent star formation. M106 appears to have a central massive object and one of the most dense nuclei known. Also the nucleus appears to be active, as it emits a lot of radio waves and the overall radio diameter of M106 suggests something strange. M106 is circumpolar from the UK, much of northern Europe and from Canada and some of the northern USA.

Where to Find

Although, technically, M106 lies within the borders of Canes Venatici, it lies in "no man's land" between the main asterisms of its host constellation and the Plough. It is best found by scanning between Gamma Ursae Majoris and Cor Caroli, although it is too faint for most finderscopes (Fig. 7.40).

Fig. 7.39 M106 by Anthony Glover

Object Type	Spiral galaxy
Declination	47° 18 min
Right Ascension	12 h 19.0 min
Magnitude	8.4

What It Looks Like

M106 is relatively bright but, being quite large, its light is spread out over quite a large area. The first view revealed a rather surprisingly large size and it reminded me of the Pinwheel (M33) as seen through binoculars. The spiral arms appeared to blink in and out of view and sometimes appeared detached from the nucleus.

How to View

Whilst a large aperture is an advantage for M106, a 127 mm Maksutov, armed with a light pollution reduction filter, is enough to show its structure. Low magnification and a wide field of view help, too.

Photographic Details

Figures 7.39 and 7.41

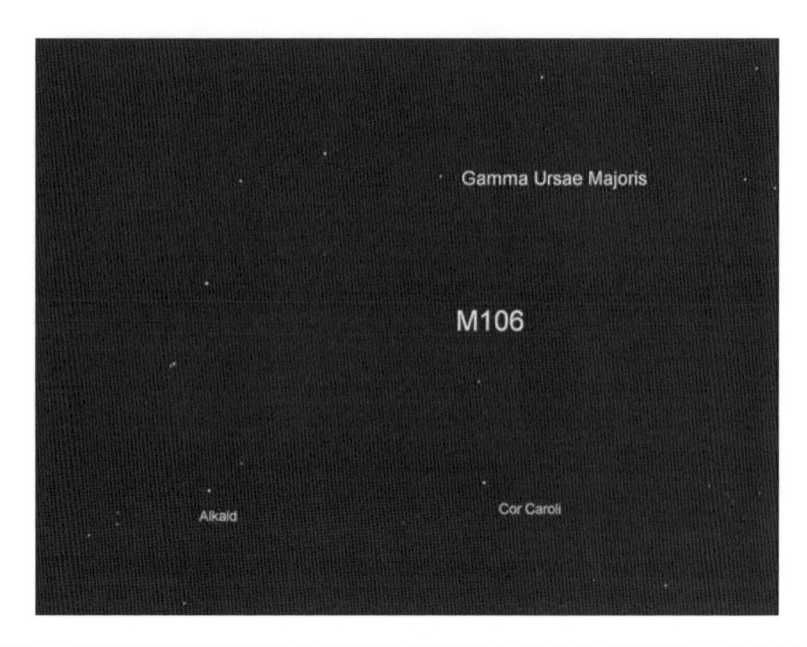

Fig. 7.40 How to find M106

Fig. 7.41 M106 modified (NASA/courtesy of nasaimages.org)

M107

M107 is a globular star cluster in Ophiuchus. It was discovered by Pierre Méchain in 1782. It is 13 arcminutes across and just under 21,000 light years away.

Where to Find

M107 is a difficult object in the south west corner of Ophiuchus, near the Scorpius border. It is about 13° due north of Antares but is too faint to detect in finderscopes (Fig. 7.42).

Object Type	Globular star cluster
Declination	−13° 03 min
Right Ascension	16 h 32.5 min
Magnitude	7.9

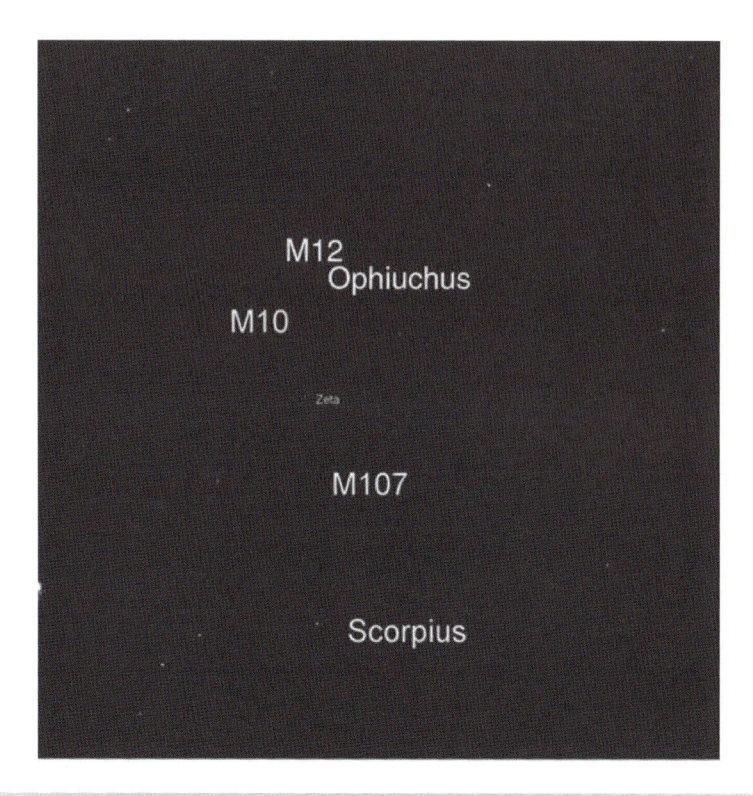

Fig. 7.42 How to find M107

What It Looks Like

M107 looks like a fuzzy patch to 70 mm binoculars. Although it is faint, a 127 mm Maksutov shows a definite central condensation of stars and diffuse outer regions. This could be confused with an elliptical galaxy.

How to View

Although M107 shows some structure in a 127 mm Maksutov and light pollution reduction filter, it is best observed with larger apertures.

Photographic Details

Figures 7.43 and 7.44

Fig. 7.43 M107 (NASA/courtesy of nasaimages.org)

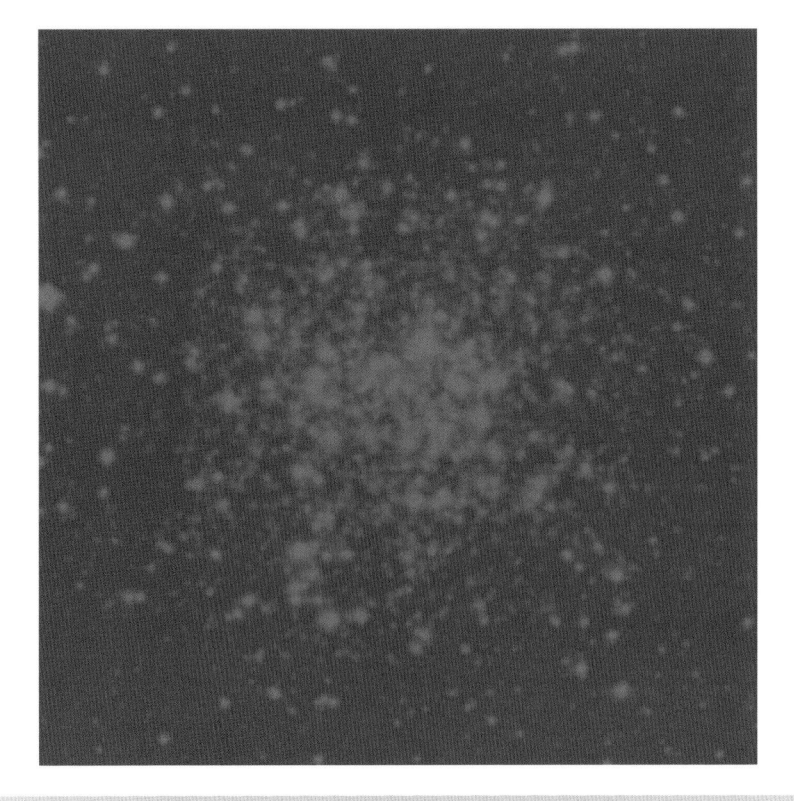

Fig. 7.44 M107 modified (NASA/courtesy of nasaimages.org)

M108

M108 is a spiral galaxy in Ursa Major. It was discovered by Pierre Méchain in 1781. It is 8×1 arcminutes in extent (suggesting an almost edge on appearance) and 45 million light years away. It is quite strange for a spiral galaxy, having only a small central bulge and a lot of dust. It is circumpolar from much of Europe and the USA.

Where to Find

M108 is very close to both Beta Ursae Majoris and the Owl Nebula (M97), although slightly nearer to M97. Given a suitably wide field of view, it is possible to see both simultaneously with a telescope. However, their faintness makes them difficult to see in a finderscope or binoculars (Fig. 7.46).

Fig. 7.45 M108 by Anthony Glover

Object Type	Spiral galaxy
Declination	55° 40 min
Right Ascension	11 h 11.5 min
Magnitude	10

What It Looks Like

M108 can be seen as a misty patch to a 127 mm Maksutov and it was impossible to determine its nature.

How to View

M108 is more an object to be crossed off the list when viewed through small aperture instruments. It needs a large aperture telescope to see properly.

Photographic Details

M108 was taken using 5×10 min exposures with an 110 mm/F7 refractor, SBIG St2000×cm, dark subtracted, stacked and processed in *Nebulosity* (Fig. 7.45).

The visual view just shows M108 as a misty patch of indeterminate object type (Fig. 7.47).

M108 345

Fig. 7.46 How to find M108

Fig. 7.47 M108 modified

M109

M109 is a spiral galaxy in Ursa Major. It was discovered by Pierre Méchain in 1781. It is 7×4 arcminutes in extent and 55 million light years away. It has a small nucleus and long, winding spiral arms. It is circumpolar from the UK and Canada and parts of the USA and Europe.

Where to Find

M109 is less than a degree from Gamma Ursae Majoris and slightly south of it. This means that it can fit into the same telescope field of view. However, its faintness makes it hard to see in finderscopes, although it can be glimpsed in larger binoculars on a clear night (Fig. 7.49).

Object Type	Spiral galaxy
Declination	53° 23 min
Right Ascension	11 h 57.6 min
Magnitude	9.8

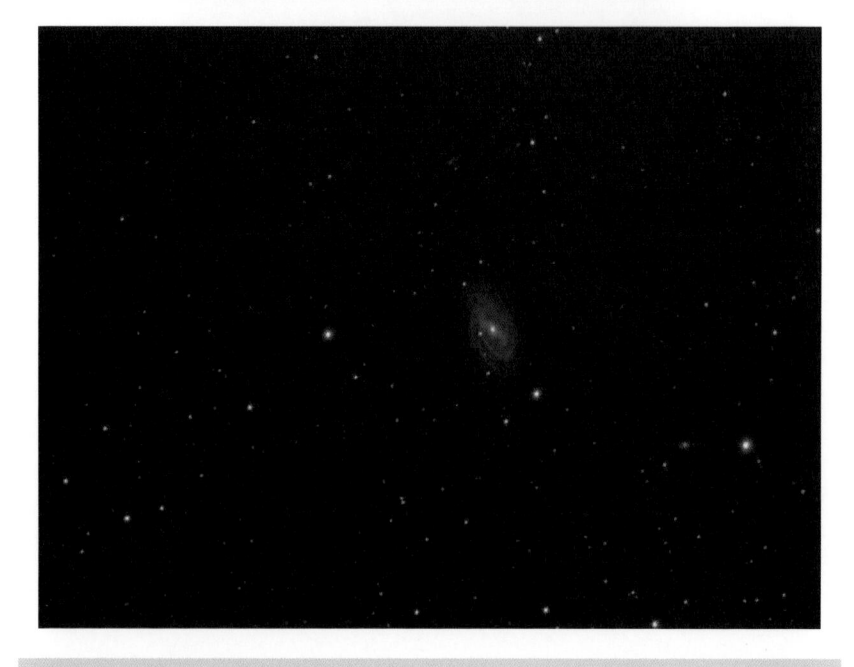

Fig. 7.48 M109 by Anthony Glover

What It Looks Like

M109 is faint and on the borderline of resolution with a 127 mm Maksutov. There was some hint of a spiral structure amongst the "fuzziness" and its position near the zenith and the light pollution reduction filter helped. Certainly, there was an obvious nucleus. However, M109 was so faint that I wasn't totally sure I'd resolved it properly.

How to View

M109 is just on the limit of resolution with my equipment, so it is preferable to use an instrument with an aperture of at least 150 mm.

Photographic Details

Figures 7.48 and 7.50

Fig. 7.49 How to find M109

Fig. 7.50 M109 modified (NASA/courtesy of nasaimages.org)

M110

M110 is an elliptical galaxy in Andromeda. It was discovered by Charles Messier (although it wasn't listed in the original list of 103 objects) in 1773 and is a satellite galaxy of the Andromeda Galaxy (M31). It is 17×10 arcminutes across and about 2.9 million light years away. It appears to have some dust clouds, not usual in this type of galaxy. Like M31, it is circumpolar from the UK and parts of Canada.

Where to Find

As M110 is a satellite galaxy of M31, it is easier to find M31 first using Beta and Mu Andromedae and pointers to it (Fig. 7.52).

Object Type	Elliptical galaxy
Declination	41° 41 min
Right Ascension	0 h 40.4 min
Magnitude	8.5

M110   349

Fig. 7.51 M110 by Anthony Glover

What It Looks Like

M110 is invisible to 70 mm binoculars and is difficult with a 127 mm Maksutov. Its appearance is not dissimilar (in small amateur instruments) to a globular star cluster and does not show any features. It is more one to cross off the list than enjoy. Despite being brighter than many Messier objects, it is not visible under poor conditions.

How to View

M110 is actually a very hard object because it is very close to the core of M31. Apart from the usual advice about aperture, it needs medium to high magnification (30-100x) to separate it from the central regions of M31.

Photographic Details

M110 is shown in the widefield shot in Fig. 4.23. Figure 4.26 shows how it looks through the 127 mm Maksutov, with a field of view of about 2° and showing some of M31 (Fig. 7.51).

Fig. 7.52 How to find M110

Acknowledgements Claudine Pouret, French Academy of Sciences.

References

Book: Jones, Kenneth Glyn; *Messier's Nebulae and Star Clusters (2nd edition);* Cambridge University Press; ISBN: 978-0521058490
Book: Crump, Thomas; *A Brief History of Science;* Running Press; ISBN: 978-0786710393
Book: Gribbin, John; *The Fellowship: The Story of a Revolution;* Allen Lane; ISBN: 978-0713997453
Book (c/o Academie de Sciences): Koertge, Noretta; *New Dictionary of Scientific Biography;* Scribner; ISBN: 978-0684313207
Website: www.seds.org/messier/xtra/history; Frommert, Hartmut and Kronberg, Christine. Last accessed: 25 March 2010
Website: www.messiermarathon.com/about.htm; [unknown]; Last accessed: 25 March 2010

Historic References

Archive: Delambre, Jean-Baptiste Joseph; *Funérailles de M. Messier;* Paris, 1817. Academie de Sciences.
Archive: Messier, Charles, & Méchain, Pierre; *Catalogue des Nébuleuses et des Amas D'étoiles;; Connaissance de Temps 1784;* Paris, 1781.

Glossary

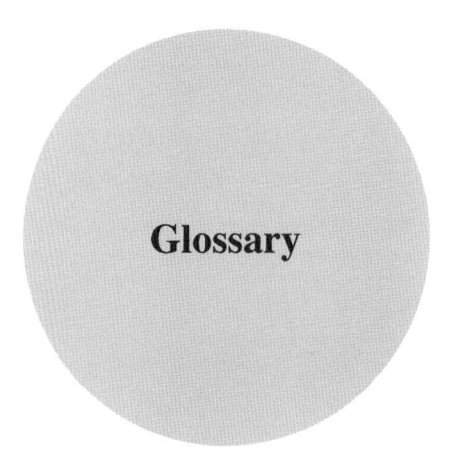

This section contains a description of some of the technical terms used in this book

Numerics

A

Abbe Nicholas Louis de la Caille   Abbe Nicholas Louis de la Caille was a French astronomer born in 1713 and he is best known for his mapping of the stars of the southern hemisphere. He introduced 14 new constellations in the far southern sky, not visible in Europe.

Abraham Ihle   Abraham Ihle was a German amateur astronomer born in 1627, who was credited with the discovery of the first globular star cluster, M22.

Absolute Magnitude   This is the magnitude that an object would appear '...if it were 10 parsecs away (the standard lumiosity distance). It is a measure of the real brightness of an object. As an example, the Sun has an absolute magnitude of 4.8.

Accessories   This is a general term used for anything that can be added to a telescope that will enable it to be used visually or photographically or change its behavior in some way. These include eyepieces, image amplifiers and focal reducers. There are all sorts of filters (q.v.) as well. There are also accessories used to improve the performance of a telescope when used for photographic use.

Achromat   See achromatic refractor (q.v.).

Achromatic Refractor This is the normal type of refractor on sale in shops. It uses an objective lens made of two types of glass (crown and flint) to reduce chromatic aberration inherent in telescope lenses.

Actual Field of View This is the diameter of the area of sky that you can see through a telescope or binoculars and is measured in degrees and/or arcminutes. It is obtained by dividing the *apparent field of view* (q.v.) of an eyepiece by the magnification.

Afocal Adaptor Afocal photography is also known as "eyepiece projection" and is the act of taking a photograph through a telescope using an eyepiece. An adaptor allows the camera to be held near the eyepiece to reduce camera shake.

Afocal Coupling This is the act of taking photographs through a telescope eyepiece.

Afocal Photography See *Afocal Coupling*, above.

Afocal Projection See *Afocal Coupling*, above.

Age of the Universe This is the estimated age of the universe from the Big Bang (q.v.). According to the current rate of expansion of the universe, extrapolated back to a time when everything converged gives 13,700 million years. This gives rise to an anomaly where some globular star clusters surrounding our Milky Way are believed to be older than that. None of this was known to Charles Messier and it doesn't affect your ability to enjoy the objects in the Messier Catalogue but it sure makes the universe seem strange.

Al-Sufi Abd-al-Rahman Al Sufi was born in ancient Persia in 903 A.D. He was working in astronomical studies in the ruler's court and continued to research Ptolemy's work. He catalogued several astronomical objects, including the Andromeda Galaxy (M31), although his work was unknown in Europe until after the invention of the telescope.

Albireo Albireo (Beta Cygni) is a well-known double star in Cygnus. As well as being known in its own right, it can also be used as a signpost to find the globular star cluster M56 in Lyra.

Alcyone Alcyone is the brightest member of the Pleiades open star cluster (M45).

Algol Algol is the well-known eclipsing binary star in Perseus and can also be used as a signpost to the open star cluster (M34).

Alkaid Alkaid is the end star of the handle of the Big Dipper or Plough asterism in Ursa Major and is used as a guide to find many deep sky objects in the Messier Catalogue.

Altair Altair is the brightest star in the constellation of Aquila. Although Aquila does not host any Messier Catalogue objects, it can be used to locate some of them.

Andromeda This is a constellation in the northern hemisphere, which is home to the famous Andromeda Galaxy (q.v.).

Andromeda Galaxy This galaxy is well known in science fiction. It is the nearest large galaxy to our own Milky Way and is the largest natural object visible in the sky from the northern hemisphere. It can be seen with the unaided eye on a good night from a dark location but normally needs binoculars.

Ångström Unit This is what you get if you divide a nanometer by 10, or a meter by 10 exactly ten times. It is used to measure the wavelength of light.

Antares Antares literally means "Rival of Mars". It is a red giant star in Scorpius and can be used to find some of the objects in the Messier Catalogue. It is also the brand name of some astronomical equipment.

Antoine Darquier de Pellepoix Darquier was born in Toulouse in 1718 where he worked as an astronomer. He is best known for discovering the Ring Nebula (M57), although its exact nature was not known until many years later.

Aperture Fever This is the obsession of many deep sky specialists who believe that the larger the aperture the better. For many objects, they are right but, on the other hand, some objects are better enjoyed with small instruments with a wide field of view.

APO/Apochromat See apochromatic refractor (q.v.).

Apochromatic Refractor This is an improvement on the achromatic refractor (q.v.). It uses three types of glass to just about eliminate chromatic aberration entirely. Care must be taken to ensure that a telescope marketed as apochromatic is not extra dispersion (ED) or semi-apochromatic. Also, there are several types of apochromatic telescope, so you need to be careful before you buy, especially if the manufacturer does not have an established track record.

Apparent Field of View This is a measure of the width of the field of view of an eyepiece. It is the theoretical field of view that would be obtained if you were able to use a magnification of 1x. To get the "real" or *actual field of view* (q.v.) for a given telescope/eyepiece combination, you need to divide the apparent field of view by the magnification. Whereas most eyepiece types have an apparent field of view of 52°, this can vary.

Apparent Magnitude This is the magnitude that an object appears in the sky, as seen from Earth. The Sun has an apparent magnitude of −26.7, Sirius has an apparent magnitude of −1.4 and the Andromeda Galaxy (M31) has an apparent magnitude of 3.7.

Aquarius Aquarius is one of the original members of the Zodiac and the Sun passes through it during late winter. It is home to some members of the Messier Catalogue.

Aratos Aratos was a Greek philosopher who noted the presence and appearance of the Pleiades (M45) and Beehive (M44).

Arcturus Arcturus is the brightest star in the constellation of Boötes and the 4th brightest star in the sky. Boötes does not contain any Messier Catalogue objects, although it can be used to find some of them.

Argo Navis Argo Navis is a large constellation which is at a very southerly declination. Only a small part of it, known as Puppis, is visible from northern Europe.

Astigmatism   This is a type of defect in optical equipment where objects near the edge of the field of view become distorted. It is most common in binoculars.

Asterism   This is a name given to a pattern of stars which is part of a constellation, rather than being a constellation in its own right. The most famous one is the Big Dipper or Plough. Some asterisms can be used to find objects in the Messier Catalogue.

Auriga   Auriga is a far northern constellation, which is home to the open star clusters M36, M37 and M38.

B

Bandpass   When using a filter, light slightly longer or shorter than the required wavelength is allowed to pass through. The range of wavelengths allowed is known as the *bandpass* and is measured in Ångström Units. A narrow bandpass is required for solar astronomy.

Barnabus Oriani   Barnabus Oriani was the director of Brera Observatory in Italy but his connection with the Messier Catalogue was the discovery of M61.

Bayer Filter Mosaic   A Bayer filter mosaic is a color filter array (CFA) for arranging RGB color filters on a square grid of photosensors. The term derives from the name of its inventor, *Bryce Bayer* of Eastman Kodak, and refers to a particular arrangement of color filters used in most single-chip digital cameras. Bryce Bayer's patent called the green photosensors *luminance-sensitive elements* and the red and blue ones *chrominance-sensitive elements*. He used twice as many green elements as red or blue to mimic the human eye's greater resolving power with green light. These elements are referred to as samples and after interpolation become pixels. The raw output of Bayer-filter cameras is referred to as a *Bayer Pattern* image. Since each pixel is filtered to record only one of the three colors, two-thirds of the color data is missing from each. A demosaicing algorithm is used to interpolate a set of complete red, green, and blue values for each point, to make an RGB image. Many different algorithms exist.

Beehive   This is a star cluster (q.v.) found in the constellation of Cancer. It is an easy target for small binoculars, and as such is one of my Usual Suspects (q.v.). It has the name M44 in the Messier Catalogue.

Big Bang   This is a set of theories that the early universe was either infinitely small and hot or finitely very small and hot (depending on exactly which version of the theory you believe in). It then started expanding. According to the popular version, time was also created at the Big Bang, so there was no "before the Big Bang", as some theories suggest.

Big Dipper   This is not a true constellation as such but is the most prominent pattern of stars (or asterism) within the constellation of Ursa Major (q.v.). It is known as the Plough in the United Kingdom.

BK-7 Type of glass used for the construction of achromatic lenses and binocular prisms.

Black Eye Galaxy This is another name for M64.

Blue Straggler In globular star clusters around the Milky Way, most of the stars are old red or yellow. There are some blue stars, which would normally indicate a younger population, hence the name. It is believed that they are caused by star mergers, which are more likely in the densely packed globular star clusters than main parts of a galaxy.

Broadband Name used to describe a filter that allows a large *bandpass* (q.v.). Broadband filters are suitable for viewing nebulae.

Brocchi's Cluster This is an alternative name for the Coathanger (q.v.).

C

Camelopardalis This is a faint far northern constellation said to represent a giraffe.

Cancer This is a constellation on the ecliptic, so forms part of the Zodiac. It also contains the Beehive (M44), one of my favorite star clusters and the similar, fainter object M67.

Carl Keenan Seyfert Carl Seyfert was a prominent academic in the United States who had researched several areas, particularly stars and galaxies. It is the Seyfert type of galaxy for which he is most remembered.

Caroline Herschel Caroline Herschel is probably better known as William's sister and John's aunt. However, she was an accomplished scientist in her own right and made several discoveries of nebulae and comets. She was born in Hanover in 1750 and came to Bath to work with William at the age of 22. Despite a childhood illness that restricted her growth to 4 ft 3ins, she lived a remarkable 98 years.

Catalogue There are several of these, which are lists of deep sky objects. The most famous is the Messier Catalogue, drawn up by Charles Messier and his colleague, Pierre Méchain.

Cassiopeia This is a far northern constellation, best known for its distinctive "W" shape. It is home to M52 and M103.

CCD/Charge Coupled Device One of the two main types of image sensors used in digital cameras. When a picture is taken, the CCD is struck by light coming through the camera's lens. Each of the thousands or millions of tiny pixels that make up the CCD converts this light into electrons. The number of electrons, usually described as the pixel's accumulated charge, is measured, and then converted to a digital value.

Centaurus A This is a galaxy that would be bright enough to make the Messier Catalogue were it not too far south for him to see it from France. It is a well-known strong radio source.

Cetus This is a southern constellation which borders onto the ecliptic. It is home to M77.

Charles Messier Charles Messier was a comet hunter in France during the late 1700s and early 1800s. However, he is better known for his "catalogue" of deep sky objects that can potentially be confused with comets. His catalogue is actually a useful list of interesting objects that are still worth looking at.

Chromatic Aberration This is an undesirable property of many telescopes and accessories which cause light of different wavelengths to come to different focal points and cause color fringes in the image. This is not a problem with narrowband solar viewing, as the light is all of the same wavelength (monochrome light, see below).

CMOS CMOS is an abbreviation of complementary metal oxide semiconductor. Pronounced see-moss, CMOS is a widely used type of semiconductor. CMOS semiconductors use both NMOS (negative polarity) and PMOS (positive polarity) circuits. Since only one of the circuit types is on at any given time, CMOS chips require less power than chips using just one type of transistor. This makes them particularly attractive for use in battery-powered devices, such as portable computers.

Coathanger This is an asterism in the constellation of Vulpecula.

Collimation This is the act of aligning the primary mirror of a reflector so that objects appear at a sharp focus.

Collinder This is a name given to a list of deep sky objects, known as a catalogue, in the same way that Messier's list of objects is known as a catalogue.

Collinder 39 This is an alternative name for Melotte 20 (q.v.).

Collinder 399 This is an alternative name for the Coathanger (q.v.).

CONR Color optimised non-raw. Mode which allows web cameras which have outputs that include image processing to be disabled.

Constellation Pattern of stars that was taken by people in the past to represent some object, animate or not. Every star is regarded as part of one of the many constellations. The convention is still used today, as it helps astronomers find their way around the sky.

Coma Berenices This is a northern constellation, most of which is actually an open star cluster in its own right, known as Melotte 111. It is home to many Messier galaxies, which are outlying members of the Virgo Supercluster.

Corrector Plate This is a glass plate used over the tube of a reflector to eliminate the effects of spherical aberration (q.v.).

Cosmic Scale This is a general term that is related to the scale of distances in the universe. The main unit of measurement is the light year, or the distance that light travels in a year. There is a lot of uncertainty in the distances of objects, particularly those outside our own galaxy. The furthest Messier object from us is about 65 million light years away from us, although it could be five million light years further or nearer. This is rather small compared to the extent of the entire universe.

Crab Nebula This is the brightest example of a supernova remnant (q.v.) and is found in Taurus but is not an easy object. It has the Messier number M1.

Cygnus Constellation thought to represent the pattern of a swan, although its alternative description of the Northern Cross is perhaps more apt. It is home to many interesting objects and the Milky Way passes through it, making it a nice place to just casually browse through binoculars. It is home to the star clusters M29 and M39.

D

Dark Nebula Perhaps one of the more obvious terms in the glossary. Appearance-wise they neither emit nor reflect light but have a similar composition to emission nebulae (q.v.) and reflection nebulae (q.v.). There are none of them in the Messier Catalogue.

Dawes Limit This is the theoretical resolution limit for a telescope. It is normally calculated by dividing the aperture in millimeters into 116 to get the answer in arc-seconds. As an example, a 114 mm Newtonian reflector has a Dawes Limit of just over an arcsecond. In practice the Dawes Limit is rarely reached. To get anywhere near it, you need a clear sky, perfect optics and a magnification of at least twice the telescope's aperture in millimetres. Yes, and a lucky charm is helpful as well!

Dumbbell This is the brightest example of a planetary nebula (q.v.) and can sometimes be seen in quite modest instruments. It is a member of the Messier Catalogue and has the number M27.

E

Eagle Nebula This is a star-forming region made famous by the Hubble Space Telescope photograph of the "Pillars of Creation" within it. It has the Messier number M17.

Ecliptic This is the path around the background sky that the Sun traces out as a result of the Earth's orbit.

ED See extra dispersion (q.v.).

Edmund Halley Most famous for discovering the existence of periodic comets, Edmund Halley was nevertheless a remarkable all-round astronomer, who discovered many properties of the Messier Catalogue objects. He was born in 1656.

Edward Pigott Edward Pigott was born in 1753 and discovered M64, a comet and the variability of Eta Aquilae, which was the first "Cepheid" variable discovered.

Edwin Hubble Edwin Hubble (born in 1889) probably needs little or no introduction, as he has the Hubble Space Telescope named after him. He is best known for his part in discovering that many members of the Messier Catalogue are galaxies, external to our own Milky Way.

Emission Nebula   This has a similar composition to a dark nebula (q.v.) or reflection nebula (q.v.) but has started to condense to form stars and young stars embedded within the nebula are emitting radiation that causes the nebula to glow.

Exit Pupil   This is the width of the light beam leaving the telescope or binocular eyepiece. It is derived by dividing the aperture by the magnification. If it matches the aperture of the observer's pupil, the results are good and objects appear bright. If it exceeds it, there is some light loss and the object appears dimmer. Young observers may have a pupil as wide as 8 mm in complete darkness but this reduces to around 5–6 mm for middle-aged observers at suburban locations. Looking at bright lights can seriously reduce your pupil width.

Extinction   This is nothing to do with the end of the human race but the dimming effect caused when an object is near the horizon. As it has to pass through more of Earth's atmosphere, more of its light is absorbed, so it is much dimmer when it meets the eye, binoculars or telescope.

Extra Dispersion   An extra dispersion lens is an improvement on the achromatic objective lens theme where it uses extra dispersion flint glass to improve the performance. It is not as good as a true apochromatic refractor and is sometimes referred to as "semi-apochromatic".

F

Faint Fuzzies   General class of object where the light is spread out over a large area, so the object appears fainter and is more difficult to see than a point source, such as a star, or a small object, such as a planet. Galaxies, globular star clusters and comets are all "faint fuzzies" and have the same observing challenges.

Filter   Device used to allow light of only certain wavelengths to pass through. There are full aperture filters (q.v.) and those placed at the eyepiece end of a telescope. They can be broadband (q.v.) and narrowband (q.v.). Night time filters are usually broadband and are placed at the eyepiece.

Flamsteed   John Flamsteed was the first Astronomer Royal. One of his contributions was to introduced the system of Flamsteed numbers (often referred to by Charles Messier). Stars are numbered from east to west in each constellation and include those of approximate unaided eye visibility.

Focal Reducer   This is the exact opposite of an image amplifier (q.v.). They are particularly useful with long focal length telescopes for viewing the Messier Catalogue.

G

Galactic Center This is the center of our Milky Way. It is obscured by interstellar dust but many objects, particularly globular star clusters are in its general direction. It never rises from northern Europe and Canada, as it is in the southern constellation of Sagittarius.

Galactic Nucleus This is another word for the center of a galaxy but in common usage it refers to the visible center of a galaxy external to our Milky Way. As it is much brighter than the outlying regions of a galaxy, it is often the only visible part as seen from Earth in small apertures or under poor viewing conditions.

Galactic Plane This is the plane of our Milky Way. Its significance is that most of the Messier objects inside our galaxy are near it.

GEM This is an abbreviation for German Equatorial Mount (q.v.).

German Equatorial Mount This is a commonly used design of an equatorial mount that includes coarse and fine controls used to find and track objects. Despite the name, most of them are made in China.

Giovanni Battista Hodierna Hodierna was an Italian cleric in the 17th century. Although many of his observations were not well preserved, it is almost certain that at least nine of the Messier objects were his original discoveries.

Globular Star Cluster This is a type of star cluster whose members are concentrated in a small area. To the unaided eye or small instruments they are hard to distinguish from small galaxies but a larger instrument may resolve outer members into individual stars.

Guilliame Le Gentil He was a French astronomer born in 1725. He discovered M32, M36 and M38. He went on an expedition to observe the transits of Venus from India, which was disrupted by both war and bad weather. He arrived home to find that he had been legally declared dead and his wife had remarried.

H

Halo A halo is the outer reaches of a galaxy, which are usually faint and often contain globular star clusters (q.v.).

H Alpha/Hydrogen Alpha This is an emission line in the nebulae that is associated with neutral hydrogen. It has a wavelength of 6,563 Å Units and is towards the red end of the spectrum.

Hercules Hercules is a northern constellation which is not particularly bright or remarkable but is home to two globular star clusters, M13 and M92.

Horsehead Nebula This is a dark nebula in Orion.

Hyades This is a star cluster in the constellation of Taurus.

I

Image Amplifier This is a general term for a telescope accessory, which boosts the magnification of any given eyepiece. The most common type is the Barlow lens.

J

Jean-Dominique Maraldi He was an Italian astronomer born in 1709 but came to Paris in 1727 and later discovered the globular star clusters M2 and M15. His original name was Giovanni Domenico Maraldi but is better known by the French version of his name.

Jean-Jacques d'Ortous de Mairan He was a French scientist born in 1678. His claim to astronomical fame was the discovery of M43.

Jewel Box This is an open star cluster in the Southern Cross. Although bright enough to make the Messier Catalogue, it is too far south to be visible from France.

Johann Elert Bode He was a German astronomer born in 1747 and was best known for Bode's Law, which speculates the distances of the planets from the Sun. He discovered M53, M81 and M82.

Johann Gottfried Koehler He was a German astronomer born in 1745 and discovered M59 and M60.

John Bevis John Bevis was an English astronomer who discovered the Crab Nebula (M1). He produced his own star atlas, which was published after his death.

K

L

Lagoon Nebula This is the popular name for M8, a large star forming region in Sagittarius.

Lenticular Galaxy This is a type of galaxy that is intermediate between an elliptical galaxy and spiral galaxy. It is a spiral galaxy that has lost its interstellar material, possibly as it has been all used up in star formation.

Lepus This is a rather inconspicuous constellation south of Orion. It is home to the globular star cluster M79.

Light Pollution Reduction This is a type of filter that blocks out some of the light produced by street lighting. Whilst it is true that it dims the object you are trying to look at, it is useful for viewing faint extended objects, such as galaxies that may otherwise be invisible.

Little Dumbell Nebula   This is the popular name for M76, a planetary nebula in Perseus.

Lobster Nebula   This is an alternative popular name for M17, a nebula in Sagittarius.

LER   Type of eyepiece where the image can be viewed a further distance than other designs. They are particularly useful for spectacle wearers and are suited to photography.

LPR   See Light Pollution Reduction (q.v.).

Luminance   "Luminance Layering" is a technique that has been developed independently by Dr. Kunihiko Okano and Robert Dalby. It allows astroimagers to overcome the "tricolor" hurdle. When used with RGB filters it is referred to as "Luminance Layering" or "LRGB" technique (the "L" referring to "luminance"). The technique can be used with other color filters such as CMY (cyan, magenta, yellow) filters or with film images with comparable success. The basic premise of luminance layering is that by combining an unfiltered high resolution and high S/N grayscale image with the weaker color data we can "buy back" the signal and detail lost in our filtered RGB exposures. The end result potentially should be a more aesthetically pleasing high contrast color image.

Lux value   This is the amount of visible light per square meter incident on a surface. 1 lux = 1 lumen/square meter = 0.093 footcandles.

Lyra   This is a northern constellation thought to represent the pattern of a lyre (a type of musical instrument). Despite its small size, it is home to many interesting objects and the Milky Way passes through it. It contains the Ring Nebula (M57).

M

M   The letter "M" itself is usually followed by a number, indicating that it is an object in the Messier Catalogue.

Maksutov-Cassegrain   Type of reflecting telescope designed to reduce chromatic aberration and produce high quality images. It is often known as a Maksutov, for short.

Messier   See Charles Messier (q.v.).

Monoceros   Monoceros is a faint constellation, east of Orion. It is home to the open star cluster M50.

Monochrome Light   Light that is of the same wavelength. The term "monochrome" as used for "black and white" television, is in fact used incorrectly. Hydrogen alpha and calcium K are examples of monochrome light.

N

Nagler   This is a type of eyepiece that is only available to the financially gifted or irresponsible but is very, very good or maybe even better than that!

Nanometer   This is a billionth of a meter and, like the Ångström Unit, is also used to measure very small distances, such as light wavelengths. It is commonly abbreviated to "nm". There are ten Ångström units in a nanometer.

Narrowband   This is a term used to describe filters with a low bandpass, which only allow light very close to a specific wavelength to pass through. These have little or no use for viewing objects in the Messier Catalogue.

Nebula   This is a large "cloud" consisting mainly of hydrogen but also some heavier elements.

Nicholas-Claude Fabri de Peiresc   Nicholas-Claude Fabri de Peiresc was a French lawyer who discovered the Orion Great Nebula (M42).

Northern Cross   This is an alternative name for Cygnus (q.v.).

O

Occulting Bar   This is a specialized piece of equipment used to block the light from a bright object, so that surrounding fainter objects can be seen.

Omega Centauri   This is the brightest globular star cluster (q.v.) in the night sky but is difficult to see from anywhere in the northern hemisphere and impossible from the U.K. and Canada.

Omega Nebula   This is an alternative popular name for M17, a nebula in Sagittarius.

Omicron Ceti   See Mira (q.v.).

Open Star Cluster   This is a star cluster where most of the members can be separated clearly with or without optical aid but are close enough together to make it obvious that they are physically associated. Examples include the Hyades, Pleiades and Beehive.

Ophiuchus   This constellation is on the ecliptic but is not part of the Zodiac, as known to astrologers. It is home to several globular star clusters in the Messier Catalogue.

Orion   This constellation is quite well known to non-astronomers. It is home to the Great Orion Nebula (M42).

Oswald Thomas   He was born in Romania in 1882. He is best known for finding the missing M48.

P

Perseus This is a northern constellation said to represent a hero. It is home to many interesting objects, such as the open star cluster M34.

Philippe Loys de Chéseaux He was a Swiss landowner and astronomer, born in 1718. He discovered several objects in the Messier Catalogue.

Pinwheel The Pinwheel is a popular name for the spiral galaxy M33 in Triangulum. It is a member of our Local Group of galaxies.

Pisces Pisces is a constellation that straddles both the celestial equator and ecliptic. It is home to the spiral galaxy M74.

Pixelated/Pixelation This is an output with large, coarse-looking pixels. Using too high a resolution (pixels smaller than the output device can produce) increases the file size and slows the printing of the image; furthermore, the device will be unable to reproduce the extra detail provided by the higher resolution image.

Planetary Nebula This is a type of object formed when a star of comparable mass to the Sun finishes its red giant phase and sheds its outer layers. The best examples in the Messier Catalogue are M27 and M57.

Pleiades Star cluster in the constellation of Taurus, also known as the Seven Sisters. They are known as M45 in the Messier Catalogue.

Plössl Commonly used type of eyepiece, suitable for visual use.

Post Processing This is the act of improving digital images following capture using a tool such as *Paintshop Pro, Photoshop* or similar.

Praesepe This is an alternative popular name for M44, an open star cluster in Cancer.

Primary Mirror This is a mirror used to collect light in a reflector, which performs the same task as an objective lens in a refractor.

Progenitor Star This is a general name for a star in a state before it became a supernova or planetary nebula.

Q

R

Reflection Nebula This has a similar composition to a dark nebula (q.v.) but there happens to be a nearby star or few that light the nebula up to make it visible.

RGB Literally "red, green and blue". In photographic terms, a color image is composed of red, green and blue components and the amount of red, green and blue in a camera pixel determines the shade of color that is reproduced. Digital photography post processing (q.v.) manipulates the red, green and blue outputs.

Ring Nebula The Ring (as it is often referred to by its shortened name by those in the know) is the most well-known planetary nebula, as it's the easiest to find (although not the brightest). It is also known as M57 in the Messier Catalogue.

RR Lyrae This is a type of variable star that has a known period/luminosity relationship, so can be used to estimate its distance accurately.

S

Satellite Galaxy This is a small galaxy that orbits a larger one. For example the Magellanic Clouds (q.v.) orbit the Milky Way and M110 and M32 orbit the Andromeda Galaxy (M31).

Scorpio This is a popular name for Scorpius (q.v.).

Scorpius Constellation in the southern part of the sky crossed by the Sun during late autumn. Apart from the ecliptic passing through it, it is also well known for its brightest star, Antares (q.v.). It is also close to the central part of the Milky Way. It has some Messier Catalogue objects but most of them are difficult from the UK, as they never rise far above the horizon.

Secondary Mirror This is a mirror used in most types of reflector to reflect the light from the primary mirror (q.v.) to the terminal light path, so that an object can be viewed through an eyepiece.

Semi-Apochromatic This is a term to be wary of. It can mean anything from a high quality achromatic refractor to something quite close to a true apochromatic one. Consult independent advice before buying. If it uses the term "ED" or "extra dispersion" (q.v.) it is quite good and you know what you're getting, although a true apochromat is rather better.

Seven Sisters See Pleiades (q.v.).

Seyfert Galaxy This is a general term for a galaxy that is active in radio waves and X rays, suggesting that it has an active nucleus, probably powered by a black hole.

Sleeping Beauty Galaxy This is a popular name for M64, a spiral galaxy in Coma Berenices.

Spherical Aberration This is an optical defect introduced by the use of a primary mirror with a spherical shape. It can be corrected by various means, such as a corrector plate or using a parabolic mirror instead.

Star We all know what a star is, or do we? For the purposes of amateur astronomy, it doesn't really make a difference. At the smallest end, the boundary between a star and a brown dwarf (the next smallest object) is a bit blurred, as is the boundary between a brown dwarf and planet.

Star Atlas This is a book or website used to locate stars and other deep sky objects.

Glossary 367

Star Cluster Group of stars which are bound by gravity and appear close together in the sky as seen from Earth.

Sunflower Galaxy This is a popular name for M77, a spiral galaxy in Cetus.

Swan Nebula This is a popular name for M17, a nebula in Sagittarius.

T

Taurus This is a constellation apparently representing a bull that is on the ecliptic and is therefore part of the Zodiac. It is home to the Crab Nebula (M1) and Pleiades (M45).

Triangulum This is a constellation between Aries and Andromeda. Its main claim to fame is M33, a spiral galaxy in our Local Group.

U

Ursa Major This is a constellation circumpolar from the United Kingdom, Canada and many parts of Europe and the United States. It is home to the Big or Plough asterism.

Usual Suspects This is a nickname given to a group of easy deep sky objects that can be seen by inexperienced observers using modest equipment. Many of them are in the Messier Catalogue. Most astronomy writers have their own list and I'm no exception.

V

Vulpecula This is a small northern constellation bordering on Cygnus (amongst others). It is home to the Coathanger (q.v.) and Dumbbell (M27).

W

Webcam Abbreviated form of "web camera", which gives a high quality, low cost means of astrophotography.

Whirlpool Galaxy This is a popular name for M51, a spiral galaxy in the constellation of Canes Venatici.

Wild Duck Cluster This is a popular name for M11, an open star cluster in the constellation of Scutum.

William Herschel William Herschel probably needs little or no introduction and is best known for his discovery of Uranus. His connection with the Messier Catalogue and deep sky viewing is the New General Catalogue (NGC), which contains all of the Messier objects and many more fainter ones.

Y

Z

Zodiac This literally means "circle of animals". It is the set of constellations (q.v.) on the ecliptic (q.v.). However, there is a lot of confusion because the zodiac as used by astrologers is about 2000 years out of date! This is one reason I am cynical about astrology. The true zodiac contains Ophiuchus and the planets and Moon can wander into Cetus and Orion as well.

Index

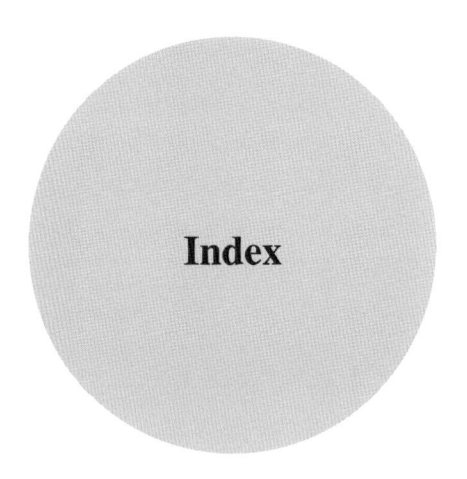

80mm APO Refractor, 127, 174, 179, 308
80mm short tube refractor, 26, 64, 135, 162

A
Accessory, 27, 105, 132, 162
Achromat, 185, 255
Achromatic refractor, 191
Active galactic nucleus, 256
Actual field of view, 354, 355
Afocal adaptor, 354
Afocal coupling, 354
Afocal photography, 354
Afocal projection, 156, 162
Age of the universe, 37, 68
Albireo, 196
Alcyone, 165
Algol, 134
Alkaid, 215
Al-Sufi, 124
Altair, 63, 78, 82, 110, 239
Altitude, 65, 72, 75, 153, 257, 335
Ambient temperature, 39, 70, 135, 140, 156, 162–163, 179
Andromeda, 9, 10, 20, 83, 124, 126, 129, 135, 255, 348
Andromeda Galaxy, 26, 27, 124, 130, 222, 224, 285, 348
Ångström Unit, 355

Antinous, 63, 111
Aperture, 6, 8, 27, 33, 36, 39, 45, 46, 49, 52, 59, 61, 63, 64, 70, 72, 73, 76, 79, 83, 86, 87, 89, 90, 92, 95, 96, 98, 103, 105, 109, 111, 114, 117, 120, 123, 125, 127, 129, 132, 135, 139, 142, 144, 146, 149, 153, 156, 159, 161, 165, 169, 171, 174, 176, 179, 181, 182, 188, 191, 194, 195, 197, 199, 200, 202, 205, 208, 211, 214, 217–219, 222, 225, 227, 230, 235, 238, 241, 244, 247, 249, 252, 254, 255, 257, 258, 261, 264, 267, 268, 270, 276, 279, 282, 284, 287, 289, 292, 295, 298, 303, 305, 308, 311, 314, 316, 319, 322, 327, 330, 335, 337, 339, 342, 344, 347, 349
  fever, 186, 333
APO, 33, 39, 114, 120, 127, 142, 144, 146, 171, 174, 179, 197, 241, 303, 308
Apochromatic, 127
  refractor, 127
  telescope, 127
Apparent field of view, 354, 355
Apparent magnitude, 112, 124
Apparent size, 31, 36, 57, 68, 135, 165, 188, 198, 201, 209, 218, 242, 256, 312, 337
Aquarius, 12, 34, 35, 63, 242, 245

369

Index

Aquila, 63, 239
Aratos, 159
Arcminute, 31, 32, 34, 36, 37, 41, 44, 46, 47,
    53, 57, 60, 61, 63, 65, 66, 68, 72,
    75, 78, 85, 87, 90, 94, 101, 106,
    108, 109, 112, 114, 115, 118, 121,
    128, 130, 134, 138, 139, 141–147,
    152, 155, 158, 159, 167, 169–172,
    174, 175, 178–180, 182, 184,
    187–189, 192, 195, 198, 201, 204,
    207, 209, 212, 215, 218, 221, 224,
    226, 228, 233, 236, 239, 242, 248,
    250, 253, 256, 259, 265, 268, 272,
    275, 277, 280, 283, 285, 286, 288,
    291, 294, 297, 301, 304, 305, 307,
    310, 312, 315, 318, 321, 323, 326,
    328, 331, 333, 334, 336, 338, 341,
    343, 346, 348
Arcturus, 38, 187
Argo Navis, 38, 187
Aries, 248
Aristotle, 147, 152
Asterism, 27, 68, 103, 120, 150, 160, 165,
    171, 245, 246, 254, 304, 338
Astigmatism, 356
Astrologer, 364, 368
Astrology, 22
Astronomik LPR filter, 39, 174, 179, 308
Auriga, 141–146

B

Bandpass, 356, 357, 364
Barlow lens, 362
Barnabus Oriani, 209
Barred spiral galaxy, 201, 202, 310
Bayer filter mosaic, 356
Beehive, 25, 135, 152, 159, 226, 227
Betelguese, 260
Bevis, J., 10, 31
Big Bang, 354, 356
Big Dipper, 165
Binoculars
    BK-7, 357
    large aperture, 61, 64, 86, 89, 146, 174,
        179, 186, 227, 247, 264, 276,
        305, 339
Black Eye Galaxy, 218
Black hole, 75, 128, 207, 256, 277, 336
Blue stragglers, 226, 265
Bode, J.E., 172, 187, 268, 272, 301
Book, 6, 26, 97, 151
Boötes, 39, 68, 326
Brera Observatory, 356
Broadband, 357, 360

Brocchi's Cluster, 357
Brown dwarf, 96, 163

C

Calcium K, 363
Camelopardalis, 23
Cancer, 160, 161, 226
Canes Venatici, 37–39, 180, 182, 215, 216,
    218, 307, 308, 338
Canis Major, 14, 152, 153, 167, 169, 179,
    304, 305
Canis Major Dwarf Galaxy, 263
Capricorn, 121, 123, 192, 242, 244, 245, 250,
    252
Cassini, 13, 20, 178
Cassiopeia, 23, 184–186, 254, 331, 332
Castor, 139
Catalogue, 13, 14, 16, 17, 19, 22–29, 32, 43,
    44, 46, 53, 58, 68, 83, 97, 101, 104,
    125, 130, 146, 164, 181, 188, 226,
    235, 245, 253, 275, 276, 279, 282,
    297, 298, 301, 303, 310, 321,
    326–328, 330, 332
CCD. *See* Charge coupled device
Celestial equator, 34, 60, 65, 68, 72, 112,
    256, 260
Centaurus, 276
Centaurus A, 275
Central massive object, 256, 336, 338
Cepheid variable, 124, 125
Cetus, 256, 258
CFA. *See* Color filter array
Charge coupled device (CCD), 114, 120,
    142, 144, 146
Chromatic aberration, 354, 355, 358, 363
Circumpolar, 68, 118, 141, 147, 149, 180, 184,
    215, 253, 269, 272, 301, 315, 326,
    331, 338, 343, 346, 348
Cloud, 10, 20, 26, 31, 35, 76, 105, 148, 156,
    163, 165, 218, 226, 348
Clouds of Magellan, 125
Cluster, 12, 25, 34, 101, 167, 233, 301
CMOS. *See* Complementary metal oxide
    semiconductor
Coathanger, 357, 358, 367
Collimation, 358
Collinder, 358
Collision, 265
Color filter array (CFA), 356
Color optimised non-raw (CONR), 358
Coma Berenices, 17, 187, 188, 218, 219, 280,
    282, 288, 297, 318, 319, 321, 323
Comet, 2–5, 7–25, 33, 36, 39, 46, 61, 63, 66,
    68, 70, 73, 114, 120, 123, 139, 144,

Index 371

173, 176, 179, 182, 185–186, 188, 197, 200, 202, 205, 207, 209, 211, 213, 217, 219, 225, 235, 236, 241, 267, 301, 303
Comet Hale-Bopp, 236
Compact digital camera, 27, 135, 156, 162
Companion galaxy, 125, 180, 288
Complementary metal oxide semiconductor (CMOS), 358
CONR. *See* Color optimised non-raw
Constellation, 8–10, 17, 23, 34, 38, 44, 61, 63, 68, 79, 113, 114, 123, 139, 160, 167, 178, 180, 188, 226, 239, 248, 256, 263, 298, 318, 338
Cor Caroli, 180, 215, 338
Corrector plate, 358, 366
Corvus, 229, 230
Cosmic scale, 124, 130
Cosmology, 37
Crab Nebula, 10, 31
Cygnus, 118–120, 147, 148, 196, 197

D

Dark matter, 268
Dark nebula, 90, 104, 260
Dawes Limit, 359
de Chéseaux, P.L., 41, 78, 81, 106, 138, 239
Declination, 27, 29, 32, 35, 38, 41–45, 48, 51–54, 58, 61, 63, 66, 68, 69, 73, 75, 76, 78, 81, 83, 86–89, 91, 94–97, 101, 105, 107, 110, 112, 113, 115, 116, 118, 119, 121, 122, 125, 129, 131, 134, 135, 139, 142, 144, 146, 148, 150–153, 155, 158, 160, 164, 168, 171, 173, 176, 179, 181, 184, 188–190, 192, 196, 198, 199, 202, 205, 207, 210, 212, 216, 218, 219, 221, 224, 226, 228–230, 233, 237, 240, 243, 244, 246, 248, 251, 254, 257, 260, 264, 266, 267, 269, 272, 276, 278, 280, 284, 286, 287, 289, 292, 295, 298, 302, 305, 307, 310, 313, 316, 319, 321, 322, 324, 326, 329, 332, 334, 335, 337, 339, 341, 344, 346, 348
Deep sky
  object, 24, 26, 105, 182, 198
  observing, 24, 26
  specialist, 355
  viewing, 25, 26, 182
Defect, 356, 366
de La Caille, A.N.L., 275
Deneb, 118, 147, 187, 318
de Peiresc, N.-C.F., 155

de Pellepoix, A.D., 198
Diffraction spikes, 165
Digital camera, 27, 135, 156, 162
Dobsonian reflector, 162
d'Ortous de Mairan, J.-J., 158
Double star, 26, 149–151, 196, 301
Draco, 23–24, 301, 328, 330
Dumbbell, 26, 198
Dumbbell Nebula, 112
Dust lane, 53, 55, 125, 126, 132, 158, 219, 328, 334

E

Eagle Nebula, 78, 81, 82
Earth, 13, 31, 34, 57, 90, 96, 118, 128
Earth's atmosphere, 118
Eclipsing binary star, 134
Ecliptic, 31, 34, 41, 92, 98
ED, 39, 132, 165, 182, 197, 222, 241, 270, 273, 303, 327
Elevation, 27, 52, 57, 60, 112, 153, 190, 200
Elliptical galaxy, 34, 35, 38, 42, 55, 69, 72, 75, 76, 97, 115, 125, 128, 129, 175, 176, 187, 189, 193, 196, 201, 204, 205, 207, 208, 264, 265, 267, 270, 279, 285, 286, 291, 292, 321, 326, 336, 337, 342, 348
Emission nebula, 90
Equator, 34, 43, 60, 65, 68, 72, 83, 112, 125, 131, 134, 152, 153, 167, 198, 253, 256, 260, 305
Equatorial mount, 361
Equuleus, 76
Exit pupil, 360
Extinction, 27, 31, 34, 41, 43, 53, 68, 85, 90, 94, 97, 115, 118, 141, 143, 152, 191, 200, 210, 250, 253, 305
Extra dispersion, 359, 360, 366
Extra dispersion flint glass, 360
Eyepiece
  wide field of view, 156, 333

F

Faint fuzzy, 13, 105, 116, 205, 213, 273
Field of view, 26, 33, 36, 46, 49, 55, 61, 64, 66, 70, 75, 78, 82, 85, 86, 91, 103, 105, 108, 114, 119, 125–127, 129, 132, 135, 139, 144, 146, 153, 156, 161, 162, 165, 167, 169, 171, 174, 179, 182, 188, 207, 221, 224, 233, 248, 256, 278, 284, 294, 305, 315, 326, 327, 331, 333, 339, 343, 346, 350

Filter
  light pollution reduction (LPR), 26, 36, 43, 45, 46, 63, 69, 83, 86, 89, 92, 95, 98, 105, 109, 122, 135, 156, 162, 165, 181, 188, 193, 200, 205, 207, 208, 214, 216, 217, 219, 229, 244, 251, 258, 261, 270, 273, 279, 282, 284, 287, 303, 308, 316, 324, 326, 327, 335, 339, 342, 347
Finderscope, 26, 31, 45, 54, 60, 65, 75, 119–120, 127, 130, 131, 134, 141, 142, 160, 167, 180, 184, 185, 204, 215, 221, 224, 240, 242, 245, 254, 256, 260, 261, 263, 269, 275, 280, 291, 307, 310, 315, 318, 326, 328, 331, 332, 338, 341, 343, 346
Flamsteed, J., 7, 18, 23, 46, 55, 61, 89, 95, 98, 103, 108, 114, 123, 169, 188, 197, 252, 282, 308, 319
Focal length, 8, 9, 20, 25, 26, 29, 49, 61, 64, 66, 70, 105, 109, 114, 127, 132, 135, 162, 169, 171, 247, 305
Focal ratio, 61, 66, 127, 132
Focal reducer, 26, 45, 49, 70, 105, 127, 132, 135, 156, 226
Focus, 21, 329
Full aperture filter, 360

G

Galactic center, 85, 87, 115, 187, 233, 242, 263
Galactic nucleus, 125, 256
Galactic plane, 228, 263
Galaxy
  barred spiral, 201, 202, 310
  elliptical, 34, 35, 38, 42, 55, 69, 72, 75, 76, 97, 115, 125, 128, 129, 175, 176, 187, 189, 193, 196, 201, 204, 205, 207, 208, 264, 265, 267, 270, 279, 285, 286, 291, 292, 321, 326, 336, 337, 342, 348
  irregular, 144, 218, 272
  lenticular, 277, 278, 280, 281, 283, 284, 328, 329
  satellite, 128, 249, 326, 348
  spiral, 55, 125, 130, 131, 181, 201, 202, 210, 215, 216, 218, 219, 221, 224, 248, 256, 257, 268, 269, 275, 276, 288, 289, 294, 295, 297, 298, 307, 310, 312, 313, 318, 319, 321, 323, 324, 326, 334, 338, 339, 343, 344, 346

Gas, 256, 285, 291
GEM, 361
Gemini, 27, 138, 139
German Equatorial Mount, 361
Glass
  crown, 354
  flint
    extra dispersion, 360
Globular star cluster, 12, 26, 27, 34, 35, 37–39, 41, 42, 44, 45, 57, 58, 60, 61, 65, 66, 68, 69, 72, 73, 75, 76, 87, 88, 94, 96, 97, 115–117, 121, 122, 125, 128, 130, 187–190, 192, 195, 196, 207, 212–214, 226, 228, 229, 233, 236, 237, 239, 240, 242, 243, 250, 251, 263–267, 285, 301, 302, 334, 337, 341, 349
Glossary, 353–368
Great Globular Cluster in Hercules, 68

H

Halley, E., 7, 8, 10, 12, 68
Halo, 9, 128
H Alpha, 172
Hercules, 37, 68, 70, 301–303
Herschel, C., 170, 172
Hersche, W., 17–19, 22, 24, 75, 124, 130, 297
Hodierna, G.B., 47, 53, 130, 134, 141, 143, 145, 152, 170
Homer, 163
Horizon, 27, 41, 49, 65, 68, 81, 101, 106, 118, 141, 143, 159, 160, 252, 263
Horsehead Nebula, 361
Horseshoe Nebula, 81
Hubble, E., 25, 124, 125
Hubble Space Telescope, 12, 78
Huygens, 13, 29, 156
Hyades, 361, 364
Hydra, 172, 226, 228, 230, 275
Hydrogen, 361, 364
Hydrogen alpha, 33, 80, 132, 182, 222, 270, 273, 327
Hydrogen alpha filter, 33, 80

I

IC 4307, 81
IC 4703, 78
Ihle, A., 96, 98
Image amplifier, 353, 360, 362
Infra red, 197, 241, 256, 272, 303
Intergalactic material, 285

Index 373

Interstellar material, 118
Irregular galaxy, 144, 218, 272

J
Jewel Box, 134, 135, 153
Jupiter, 2, 8, 11, 16, 121, 200

K
Keystone, 68
Kirch, G., 44, 63
Koehler, J.G., 204, 207, 226

L
Lagoon Nebula, 53
Le Gentil, G., 128, 147
Le Gentil, M., 29, 98, 127–129, 147
Lens
　objective, 360, 365
Lenticular galaxy, 277, 278, 280, 283, 284, 328, 329
Leo, 221–225, 310–312, 314, 318, 336, 337
Leo I Group, 310, 312, 336
Leo Triplet, 221–225
Lepus, 188, 263, 264, 308
LER, 363
Libra, 46
Light beam, 360
Light pollution, 26, 41, 316, 319, 322
Light pollution reduction (LPR) filter, 26, 36, 43, 45, 46, 63, 69, 83, 86, 89, 92, 95, 98, 105, 109, 122, 135, 156, 162, 165, 181, 188, 193, 200, 205, 207, 208, 214, 216, 217, 219, 229, 244, 251, 258, 261, 270, 273, 279, 282, 284, 287, 303, 308, 316, 324, 326, 327, 335, 339, 342, 347
Light year, 31, 34, 37, 41, 44, 47, 51, 53, 57, 60, 63, 65, 68, 72, 75, 78, 82, 85, 87, 90, 96, 101, 104, 106, 109, 112, 115, 118, 121, 124, 130, 134, 138, 141, 143, 145, 147, 149, 152, 155, 159, 163, 167, 170, 172, 175, 178, 180, 184, 187, 189, 195, 198, 201, 204, 207, 209, 212, 215, 218, 221, 224, 226, 228, 233, 236, 239, 242, 245, 248, 250, 253, 256, 259, 263, 265, 268, 272, 275, 277, 280, 285, 288, 291, 294, 297, 301, 304, 307, 310, 312, 315, 318, 321, 323, 326, 328, 331, 334, 336, 338, 341, 343, 346, 348

Little Dumbell Nebula, 253
Lobster Nebula, 81
Local Group, 17, 124, 125, 130, 175
Long eye relief, 21
LPR filter. *See* Light pollution reduction filter
Luminance, 39, 132, 182, 222, 270, 273, 327
Lux value, 363
Lyra, 68, 113, 196

M
M1
　Charles Messier's original notes, 32–33
　how to view, 33
　photographic details, 33–34
　what it looks like, 32
　where to find
　　declination, 32
　　magnitude, 32
　　object type, 32
　　right ascension, 32
M2
　Charles Messier's original notes, 35–36
　how to view, 36
　photographic details, 36–37
　what it looks like, 35
　where to find
　　declination, 35
　　magnitude, 35
　　object type, 35
　　right ascension, 35
M3
　Charles Messier's original notes, 39
　how to view, 39
　photographic details, 39–41
　what it looks like, 38–39
　where to find
　　declination, 38
　　magnitude, 38
　　object Type, 38
　　right ascension, 38
M4
　Charles Messier's original notes, 42–43
　how to view, 43
　photographic details, 43–44
　what it looks like, 42
　where to find
　　declination, 42
　　magnitude, 42
　　object Type, 42
　　right ascension, 42

M5
  Charles Messier's original notes, 46
  how to view, 46
  photographic details, 46, 47
  what it looks like, 45
  where to find
    declination, 45
    magnitude, 45
    object type, 45
    right ascension, 45
M6
  Charles Messier's original notes, 49
  how to view, 49
  photographic details, 49–50
  what it looks like, 49
  where to find
    declination, 48
    magnitude, 48
    object type, 48
    right ascension, 48
M7
  Charles Messier's original notes, 52
  how to view, 52
  photographic details, 52–53
  what it looks like, 52
  where to find
    declination, 52
    magnitude, 52
    object type, 52
    right ascension, 52
M8
  Charles Messier's original notes, 55
  how to view, 55
  photographic details, 55–56
  what it looks like, 55
  where to find
    declination, 54
    magnitude, 54
    object type, 54
    right ascension, 54
M9
  Charles Messier's original notes, 59
  how to view, 59
  photographic details, 59
  what it looks like, 58
  where to find
    declination, 58
    magnitude, 58
    object type, 58
    right ascension, 58
M10
  Charles Messier's original notes, 61
  how to view, 61
  photographic details, 61–62
  what it looks like, 61
  where to find
    declination, 61
    magnitude, 61
    object type, 61
    right ascension, 61
M11
  Charles Messier's original notes, 63–64
  how to view, 64
  photographic details, 64
  what it looks like, 63
  where to find
    declination, 63
    magnitude, 63
    object type, 63
    right ascension, 63
M12
  Charles Messier's original notes, 66
  how to view, 66
  photographic details, 66–67
  what it looks like, 66
  where to find
    declination, 66
    magnitude, 66
    object type, 66
    right ascension, 66
M13
  Charles Messier's original notes, 70
  how to view, 70
  photographic details, 70–71
  what it looks like, 69
  where to find
    declination, 69
    magnitude, 69
    object type, 69
    right ascension, 69
M14
  Charles Messier's original notes, 73
  how to view, 73
  photographic details, 73–74
  what it looks like, 73
  where to find
    declination, 73
    magnitude, 73
    object type, 73
    right ascension, 73
M15
  Charles Messier's original notes, 76
  how to view, 76
  photographic details, 76–77
  what it looks like, 76
  where to find
    declination, 76
    magnitude, 76
    object type, 76
    right ascension, 76

Index

M16
  Charles Messier's original notes, 79
  how to view, 79
  photographic details, 80–81
  what it looks like, 78
  where to find
    declination, 78
    magnitude, 78
    object type, 78
    right ascension, 78
M17
  Charles Messier's original notes, 83
  how to view, 83
  photographic details, 83–84
  what it looks like, 83
  where to find
    declination, 83
    magnitude, 83
    object type, 83
    right ascension, 83
M18
  Charles Messier's original notes, 86
  how to view, 86
  photographic details, 86–87
  what it looks like, 86
  where to find
    declination, 86
    magnitude, 86
    object type, 86
    right ascension, 86
M19
  Charles Messier's original notes, 89
  how to view, 89
  photographic details, 89–90
  what it looks like, 88
  where to find
    declination, 88
    magnitude, 88
    object Type, 88
    right ascension, 88
M20
  Charles Messier's original notes, 92
  how to view, 92
  photographic details, 92–93
  what it looks like, 92
  where to find
    declination, 91
    magnitude, 91
    object type, 91
    right ascension, 91
M21
  Charles Messier's original notes, 95
  how to view, 95
  photographic details, 95–96
  what it looks like, 95
  where to find
    declination, 95
    magnitude, 95
    object Type, 95
    right ascension, 95
M22
  Charles Messier's original notes, 98
  how to view, 98
  photographic details, 98–99
  what it looks like, 97
  where to find
    declination, 97
    magnitude, 97
    object type, 97
    right ascension, 97
M23
  Charles Messier's original notes, 102–103
  how to view, 103
  photographic details, 103
  what it looks like, 102
  where to find
    declination, 101
    magnitude, 101
    object type, 101
    right ascension, 101
M24
  Charles Messier's original notes, 105
  how to view, 105
  photographic details, 105–106
  what it looks like, 105
  where to find
    declination, 105
    magnitude, 105
    object Type, 105
    right ascension, 105
M25
  Charles Messier's original notes, 108
  how to view, 108–109
  photographic details, 109
  what it looks like, 107
  where to find
    declination, 107
    magnitude, 107
    object type, 107
    right ascension, 107
M26
  Charles Messier's original notes, 111
  how to view, 111
  photographic details, 111
  what it looks like, 110–111
  where to find
    declination, 110
    magnitude, 110
    object type, 110
    right ascension, 110

M27
  Charles Messier's original notes, 114
  how to view, 114
  photographic details, 114–115
  what it looks like, 113–114
  where to find
    declination, 113
    magnitude, 113
    object Type, 113
    right ascension, 113
M28
  Charles Messier's original notes, 116–117
  how to view, 117
  photographic details, 117–118
  what it looks like, 116
  where to find
    declination, 116
    magnitude, 116
    object type, 116
    right ascension, 116
M29
  Charles Messier's original notes, 120
  how to view, 120
  photographic details, 120–121
  what it looks like, 119–120
  where to find
    declination, 119
    magnitude, 119
    object type, 119
    right ascension, 119
M30
  Charles Messier's original notes, 122–123
  how to view, 123
  photographic details, 123–124
  what it looks like, 122
  where to find
    declination, 122
    magnitude, 122
    object type, 122
    right ascension, 122
M31
  Charles Messier's original notes, 126–127
  how to view, 127
  photographic details, 127–128
  what it looks like, 125–126
  where to find
    declination, 125
    magnitude, 125
    object type, 125
    right ascension, 125
M32
  Charles Messier's original notes, 129
  how to view, 129
  photographic details, 129–130
  what it looks like, 129
  where to find
    declination, 129
    magnitude, 129
    object Type, 129
    right ascension, 129
M33
  Charles Messier's original notes, 132
  how to view, 132
  photographic details, 132–133
  what it looks like, 132
  where to find
    declination, 131
    magnitude, 131
    object type, 131
    right ascension, 131
M34
  Charles Messier's original notes, 135
  how to view, 135
  photographic details, 135–137
  what it looks like, 135
  where to find
    declination, 134
    magnitude, 134
    object Type, 134
    right ascension, 134
M35
  Charles Messier's original notes, 139
  how to view, 139
  photographic details, 140
  what it looks like, 139
  where to find
    declination, 139
    magnitude, 139
    object type, 139
    right ascension, 139
M36
  Charles Messier's original notes, 142
  how to view, 142
  photographic details, 142–143
  what it looks like, 142
  where to find
    declination, 142
    magnitude, 142
    object type, 142
    right ascension, 142
M37
  Charles Messier's original notes, 144
  how to view, 144
  photographic details, 144–145
  what it looks like, 144
  where to find
    declination, 144
    magnitude, 144

Index 377

    object Type, 144
    right ascension, 144
M38
  Charles Messier's original notes, 146
  how to view, 146
  photographic details, 146–147
  what it looks like, 146
  where to find
    declination, 146
    magnitude, 146
    object type, 146
    right ascension, 146
M39
  Charles Messier's original notes, 148
  how to view, 149
  photographic details, 149
  what it looks like, 148
  where to find
    declination, 148
    magnitude, 148
    object type, 148
    right ascension, 148
M40
  Charles Messier's original notes, 151
  how to view, 151
  photographic details, 151
  what it looks like, 150
  where to find
    declination, 150
    magnitude, 150
    object type, 150
    right ascension, 150
M41
  Charles Messier's original notes, 153
  how to view, 153
  photographic details, 153–154
  what it looks like, 153
  where to find
    declination, 153
    magnitude, 153
    object Type, 153
    right ascension, 153
M42
  Charles Messier's original notes, 156
  how to view, 156
  photographic details, 156–158
  what it looks like, 156
  where to find
    declination, 155
    magnitude, 155
    object type, 155
    right ascension, 155
M43
  Charles Messier's original notes, 159
  how to view, 159
  photographic details, 159
  what it looks like, 159
  where to find
    declination, 158
    magnitude, 158
    object Type, 158
    right ascension, 158
M44
  Charles Messier's original notes, 161
  how to view, 162
  photographic details, 162–163
  what it looks like, 160–161
  where to find
    declination, 160
    magnitude, 160
    object type, 160
    right ascension, 160
M45
  Charles Messier's original notes, 165
  how to view, 165
  photographic details, 165–166
  what it looks like, 165
  where to find
    declination, 164
    magnitude, 164
    object Type, 164
    right ascension, 164
M46
  Charles Messier's original notes, 168–169
  how to view, 169
  photographic details, 169–170
  what it looks like, 168
  where to find
    declination, 168
    magnitude, 168
    object type, 168
    right ascension, 168
M47
  Charles Messier's original notes, 171
  how to view, 171
  photographic details, 171–172
  what it looks like, 171
  where to find
    declination, 171
    magnitude, 171
    object type, 171
    right ascension, 171
M48
  Charles Messier's original notes, 173–174
  how to view, 174
  photographic details, 174
  what it looks like, 173

M48 (*cont.*)
  where to find
    declination, 173
    magnitude, 173
    object Type, 173
    right ascension, 173
M49
  Charles Messier's original notes, 176
  how to view, 176
  photographic details, 176–177
  what it looks like, 176
  where to find
    declination, 176
    magnitude, 176
    object type, 176
    right ascension, 176
M50
  Charles Messier's original notes, 179
  how to view, 179
  photographic details, 179–180
  what it looks like, 179
  where to find
    declination, 179
    magnitude, 179
    object type, 179
    right ascension, 179
M51
  Charles Messier's original notes, 182
  how to view, 182
  photographic details, 182–184
  what it looks like, 181
  where to find
    declination, 181
    magnitude, 181
    object type, 181
    right ascension, 181
M52
  Charles Messier's original notes, 185–186
  how to view, 186
  photographic details, 186
  what it looks like, 185
  where to find
    declination, 184
    magnitude, 184
    object Type, 184
    right ascension, 184
M53
  Charles Messier's original notes, 188
  how to view, 188
  photographic details, 188–189
  what it looks like, 188
  where to find
    declination, 188
    magnitude, 188
    object type, 188
    right ascension, 188
M54
  Charles Messier's original notes, 190–191
  how to view, 191
  photographic details, 191–192
  what it looks like, 190
  where to find
    declination, 190
    magnitude, 190
    object Type, 190
    right ascension, 190
M55
  Charles Messier's original notes, 193
  how to view, 194
  photographic details, 194–195
  what it looks like, 193
  where to find
    declination, 192
    magnitude, 192
    object type, 192
    right ascension, 192
M56
  Charles Messier's original notes, 197
  how to view, 197
  photographic details, 197–198
  what it looks like, 196
  where to find
    declination, 196
    magnitude, 196
    object Type, 196
    right ascension, 196
M57
  Charles Messier's original notes, 200
  how to view, 200
  photographic details, 200–201
  what it looks like, 199
  where to find
    declination, 199
    magnitude, 199
    object type, 199
    right ascension, 199
M58
  Charles Messier's original notes, 202
  how to view, 202
  photographic details, 202–203
  what it looks like, 202
  where to find
    declination, 202
    magnitude, 202
    object type, 202
    right ascension, 202
M59
  Charles Messier's original notes, 205

Index

how to view, 205
photographic details, 205–206
what it looks like, 205
where to find
   declination, 205
   magnitude, 205
   object Type, 205
   right ascension, 205

M60
Charles Messier's original notes, 207
how to view, 208
photographic details, 208–209
what it looks like, 207
where to find
   declination, 207
   magnitude, 207
   object type, 207
   right ascension, 207

M61
Charles Messier's original notes, 211
how to view, 211
photographic details, 211–212
what it looks like, 210
where to find
   declination, 210
   magnitude, 210
   object type, 210
   right ascension, 210

M62
Charles Messier's original notes, 213
how to view, 214
photographic details, 214–215
what it looks like, 213
where to find
   declination, 212
   magnitude, 212
   object type, 212
   right ascension, 212

M63
Charles Messier's original notes, 216–217
how to view, 217
photographic details, 217
what it looks like, 216
where to find
   declination, 216
   magnitude, 216
   object Type, 216
   right ascension, 216

M64
Charles Messier's original notes, 219
how to view, 219
photographic details, 219–220
what it looks like, 219
where to find
   declination, 219
   magnitude, 219
   object type, 219
   right ascension, 219

M65
Charles Messier's original notes, 222
how to view, 222
photographic details, 222–223
what it looks like, 222
where to find
   declination, 221
   magnitude, 221
   object Type, 221
   right ascension, 221

M66
Charles Messier's original notes, 224–225
how to view, 225
photographic details, 225
what it looks like, 224
where to find
   declination, 224
   magnitude, 224
   object type, 224
   right ascension, 224

M67
Charles Messier's original notes, 226
how to view, 227
photographic details, 227
what it looks like, 226
where to find
   declination, 226
   magnitude, 226
   object Type, 226
   right ascension, 226

M68
Charles Messier's original notes, 230
how to view, 230
photographic details, 230–231
what it looks like, 229
where to find
   declination, 229
   magnitude, 229
   object type, 229
   right ascension, 229

M69
Charles Messier's original notes, 234–235
how to view, 235
photographic details, 235–236
what it looks like, 234
where to find
   declination, 233
   magnitude, 233
   object type, 233
   right ascension, 233

M70
    Charles Messier's original notes, 237–238
    how to view, 238
    photographic details, 238–239
    what it looks like, 237
    where to find
        declination, 237
        magnitude, 238
        object type, 237
        right ascension, 237
M71
    Charles Messier's original notes, 241
    how to view, 241
    photographic details, 241–242
    what it looks like, 240
    where to find
        declination, 240
        magnitude, 240
        object type, 240
        right ascension, 240
M72
    Charles Messier's original notes, 243–244
    how to view, 244
    photographic details, 244–245
    what it looks like, 243
    where to find
        declination, 243
        magnitude, 243
        object type, 243
        right ascension, 243
M73
    Charles Messier's original notes, 246–247
    how to view, 247
    photographic details, 247
    what it looks like, 246
    where to find
        declination, 246
        magnitude, 246
        object type, 246
        right ascension, 246
M74
    Charles Messier's original notes, 249
    how to view, 249
    photographic details, 249–250
    what it looks like, 249
    where to find
        declination, 248
        magnitude, 248
        object type, 248
        right ascension, 248
M75
    Charles Messier's original notes, 251–252
    how to view, 252
    photographic details, 252–253
    what it looks like, 251
    where to find
        declination, 251
        magnitude, 251
        object type, 251
        right ascension, 251
M76
    Charles Messier's original notes, 255
    how to view, 255
    photographic details, 255–256
    what it looks like, 254
    where to find
        declination, 254
        magnitude, 254
        object type, 254
        right ascension, 254
M77
    Charles Messier's original notes, 258
    how to view, 258
    photographic details, 258–259
    what it looks like, 257
    where to find
        declination, 257
        magnitude, 257
        object type, 257
        right ascension, 257
M78
    Charles Messier's original notes, 261
    how to view, 261
    photographic details, 261–262
    what it looks like, 261
    where to find
        declination, 260
        magnitude, 260
        object type, 260
        right ascension, 260
M79
    Charles Messier's original notes, 264
    how to view, 264
    photographic details, 264–265
    what it looks like, 262
    where to find
        declination, 262
        magnitude, 262
        object type, 262
        right ascension, 262
M80
    Charles Messier's original notes, 267
    how to view, 267
    photographic details, 267–268
    what it looks like, 266
    where to find
        declination, 266
        magnitude, 266

Index

    object type, 266
    right ascension, 266
M81
    Charles Messier's original notes, 270
    how to view, 270
    photographic details, 270–272
    what it looks like, 269
    where to find
        declination, 269
        magnitude, 269
        object type, 269
        right ascension, 269
M82
    Charles Messier's original notes, 273
    how to view, 273
    photographic details, 273–274
    what it looks like, 273
    where to find
        declination, 272
        magnitude, 272
        object type, 272
        right ascension, 272
M83
    Charles Messier's original notes, 276
    how to view, 276
    photographic details, 276–277
    what it looks like, 276
    where to find
        declination, 276
        magnitude, 276
        object type, 276
        right ascension, 276
M84
    Charles Messier's original notes, 278–279
    how to view, 279
    photographic details, 279–280
    what it looks like, 278
    where to find
        declination, 278
        magnitude, 278
        object type, 278
        right ascension, 278
M85
    Charles Messier's original notes, 281–282
    how to view, 282
    photographic details, 282–283
    what it looks like, 281
    where to find
        declination, 280
        magnitude, 280
        object type, 280
        right ascension, 280

M86
    Charles Messier's original notes, 284
    how to view, 284
    photographic details, 284–285
    what it looks like, 284
    where to find
        declination, 284
        magnitude, 284
        object type, 284
        right ascension, 284
M87
    Charles Messier's original notes, 287
    how to view, 287
    photographic details, 287–288
    what it looks like, 286
    where to find
        declination, 286
        magnitude, 286
        object type, 286
        right ascension, 286
M88
    Charles Messier's original notes, 289
    how to view, 289
    photographic details, 289–291
    what it looks like, 289
    where to find
        declination, 289
        magnitude, 289
        object type, 289
        right ascension, 289
M89
    Charles Messier's original notes, 292
    how to view, 292
    photographic details, 292–293
    what it looks like, 292
    where to find
        declination, 292
        magnitude, 292
        object type, 292
        right ascension, 292
M90
    Charles Messier's original notes, 295
    how to view, 295
    photographic details, 295–297
    what it looks like, 295
    where to find
        declination, 295
        magnitude, 295
        object type, 295
        right ascension, 295
M91
    Charles Messier's original notes, 298
    how to view, 298
    photographic details, 298–299

382 Index

M91 (cont.)
 what it looks like, 298
 where to find
  declination, 298
  magnitude, 298
  object type, 298
  right ascension, 298
M92
 Charles Messier's original notes, 302–303
 how to view, 303
 photographic details, 303–304
 what it looks like, 302
 where to find
  declination, 302
  magnitude, 302
  object type, 302
  right ascension, 302
M93
 Charles Messier's original notes, 305
 how to view, 305
 photographic details, 305–306
 what it looks like, 305
 where to find
  declination, 305
  magnitude, 305
  object type, 305
  right ascension, 305
M94
 Charles Messier's original notes, 308
 how to view, 308
 photographic details, 308–309
 what it looks like, 308
 where to find
  declination, 307
  magnitude, 307
  object type, 307
  right ascension, 307
M95
 Charles Messier's original notes, 311
 how to view, 311
 photographic details, 311–312
 what it looks like, 310
 where to find
  declination, 310
  magnitude, 310
  object type, 310
  right ascension, 310
M96
 Charles Messier's original notes, 314
 how to view, 314
 photographic details, 314–315
 what it looks like, 313
 where to find
  declination, 313
  magnitude, 313
  object type, 313
  right ascension, 313
M97
 Charles Messier's original notes, 316
 how to view, 316
 photographic details, 316–317
 what it looks like, 316
 where to find
  declination, 316
  magnitude, 316
  object type, 316
  right ascension, 316
M98
 Charles Messier's original notes, 319
 how to view, 319
 photographic details, 319–320
 what it looks like, 319
 where to find
  declination, 319
  magnitude, 319
  object type, 319
  right ascension, 319
M99
 Charles Messier's original notes, 321
 how to view, 322
 photographic details, 322–323
 what it looks like, 321
 where to find
  declination, 321
  magnitude, 321
  object type, 321
  right ascension, 321
M100
 Charles Messier's original notes, 324
 how to view, 324
 photographic details, 324–325
 what it looks like, 324
 where to find
  declination, 324
  magnitude, 324
  object type, 324
  right ascension, 324
M101
 Charles Messier's original notes, 326
 how to view, 327
 photographic details, 327–328
 what it looks like, 326
 where to find
  declination, 326
  magnitude, 326
  object type, 326
  right ascension, 326

M102
  Charles Messier's original notes, 330
  how to view, 330
  photographic details, 330–331
  what it looks like, 329
  where to find
    declination, 329
    magnitude, 329
    object type, 329
    right ascension, 329
M103
  Charles Messier's original notes, 332
  how to view, 333
  photographic details, 333
  what it looks like, 332
  where to find
    declination, 332
    magnitude, 332
    object type, 332
    right ascension, 332
M104
  how to view, 335
  photographic details, 335
  what it looks like, 335
  where to find
    declination, 334
    magnitude, 334
    object type, 334
    right ascension, 334
M105
  how to view, 337
  photographic details, 337–338
  what it looks like, 337
  where to find
    declination, 337
    magnitude, 337
    object type, 337
    right ascension, 337
M106
  how to view, 339
  photographic details, 340
  what it looks like, 339
  where to find
    declination, 339
    magnitude, 339
    object type, 339
    right ascension, 339
M107
  how to view, 342
  photographic details, 342–343
  what it looks like, 342
  where to find
    declination, 341
    magnitude, 341
    object type, 341
    right ascension, 341
M108
  how to view, 344
  photographic details, 344–345
  what it looks like, 344
  where to find
    declination, 344
    magnitude, 344
    object type, 344
    right ascension, 344
M109
  how to view, 347
  photographic details, 347–348
  what it looks like, 347
  where to find
    declination, 346
    magnitude, 346
    object type, 346
    right ascension, 346
M110
  how to view, 349
  photographic details, 350
  what it looks like, 349
  where to find
    declination, 348
    magnitude, 348
    object type, 348
    right ascension, 348
Magellanic Clouds, 125
Magnification
  visual, 199
Maksutov, 36, 39, 42, 43, 45, 49, 52, 55, 58, 61, 64, 66, 69, 70, 73, 78, 83, 86, 88, 92, 95, 97, 102, 103, 105, 107, 109–111, 113, 114, 116, 117, 120, 122, 129, 132, 135, 140, 142, 144, 146, 148–151, 153, 156, 158, 162, 165, 168, 171, 173, 174, 176, 179, 181, 185, 186, 188, 190, 193, 196, 199, 202, 205, 207, 208, 210, 216, 217, 219, 225, 226, 229, 234, 237, 240, 243, 244, 246, 247, 249, 251, 254, 257, 264, 266, 273, 276, 278, 279, 281, 284, 286, 288, 289, 292, 295, 298, 302, 305, 308, 310, 316, 319, 321, 324, 327, 329, 332, 333, 335, 337, 339, 342, 344, 347, 349, 350
Maksutov-Cassegrain, 25, 26, 28, 32, 64, 76, 112, 132, 135, 139, 161, 162, 165, 198, 222, 224, 261, 264, 326
Maraldi, J.-D., 12, 75
Marius, S., 126

Mars, 31
Méchain, P., 17–19, 21, 23–24, 215, 242, 248, 250, 253, 256, 259, 263, 280, 307, 310, 312, 315, 318, 321, 323, 326, 328, 330, 332, 334, 336, 338, 341, 343, 346
Melotte, 187, 218, 280
Merger, 226, 265, 307
Messier, C., 1–25, 28, 32–33, 35–37, 39, 42–43, 46, 49, 52, 55, 57, 59–61, 63–66, 70, 72, 73, 76, 79, 81, 83, 85–87, 89, 90, 92, 94, 95, 98, 101–105, 108–118, 120–123, 125–127, 129, 130, 132, 135, 139, 144, 146–149, 151, 153, 156, 159, 171, 173–176, 178–180, 182, 184, 190–191, 197, 200, 202, 205, 209, 212, 216–217, 221, 222, 224, 226–227, 233–235, 241, 246–247, 251–252, 261, 267, 276, 278–279, 287, 289, 295, 298, 302–303, 314, 316, 319, 321, 324, 348
Messier Catalogue, 13, 17, 19, 24, 25, 27, 29, 32, 44, 53, 58, 68, 97, 101, 104, 125, 146, 164, 181, 188, 226, 245, 253, 275, 297, 301, 310, 321, 326–328
Messier number, 24
Messier object
  introduction, 25–29
Milky Way, 17, 26, 27, 41, 42, 52, 53, 57, 63, 64, 82, 86, 87, 101, 104–106, 118–120, 124, 125, 128, 130, 138, 139, 147, 149, 185, 189, 197, 207, 209, 248, 254, 263, 285, 288, 312, 326, 331, 332
Mini quasar, 256
Mintaka, 261
Mira, 364
Mirror, 8, 12, 17, 29
Mizar, 326
Molecular cloud, 163
Monoceros, 167, 169, 174, 178, 179
Monochrome light, 358, 363
Moon, 2, 10, 12, 13, 23, 27, 34, 35, 41, 42, 49, 78, 90, 96, 252
Moonlight, 26, 31

N
Nagler, 364
Nanometer, 355, 364
Narrowband, 33
Nebula, 9, 25, 31, 104, 169, 234, 302

Nebulosity, 33, 76, 86, 90, 92, 95, 105, 107, 108, 111, 135, 140, 142, 144, 146, 150, 153, 158, 159, 163, 165, 169, 171, 174, 182, 185, 186, 217, 226, 227, 245–247, 252, 253, 255, 258, 261, 264, 267, 270, 279, 288, 302–303, 305, 308, 316, 324, 344
Neptune, 25
New General Catalogue (NGC), 24
Newtonian, 39, 132, 182, 222, 270, 273, 327
Newtonian reflector, 8, 10, 25
NGC. *See* New General Catalogue
NGC2158, 138
NGC 3623, 221, 224
NGC3628, 222
NGC 5195, 180, 181
NGC 5866, 328, 330
NGC6530, 53, 55
NGC 6611, 78
Northern Cross, 112, 189
Nova, 24, 265
Nucleus, 69, 125, 128, 129, 176, 188, 216, 235, 256, 257, 261, 267, 269, 273, 281, 303, 319, 321, 324, 337–339, 346, 347

O
Object
  solar system, 31, 34, 41
Objective lens, 354, 360, 365
Occulting bar, 364
Omega Centauri, 42, 68, 73, 105
Omega Nebula, 81
Omicron ceti, 364
Omicron Virginis, 175
Open star cluster, 14, 26, 27, 47–49, 51–53, 63, 78, 85, 86, 94, 95, 101, 102, 106, 107, 109, 110, 118, 119, 132, 134, 138, 139, 141–148, 152, 153, 160, 163, 164, 167, 168, 170–173, 178, 179, 184, 187, 218, 226, 227, 246, 304, 305, 331, 332
Ophiuchus, 55, 59–61, 65, 66, 73, 89, 92, 103, 212, 341
Optical equipment, 356
Optics, 31
Orbit, 6–9, 17–19, 21
Orion, 29, 138, 156, 159, 259–261
Orion Great Nebula, 53, 78, 82, 155, 158, 259
Owl Nebula, 315, 343

Index

P
Paintshop Pro, 365
Parabolic mirror, 366
Parsec, 353
Pegasus, 75, 76
Perseus, 134, 253, 254
Perseus Double Cluster, 134
Photograph, 26–28, 33–34, 36–37, 39–41, 43–44, 47–57, 59–68, 70–87, 89–96, 98–99, 102–106, 109–115, 117–121, 123–125, 127–138, 140–147, 149–160, 162–166, 169–172, 174–180, 182–184, 186–189, 191–192, 194–212, 214–215, 217–231, 233, 235–236, 238–239, 241–242, 244–245, 247–250, 252–256, 258–265, 267–280, 282–284, 287–299, 303–328, 330–331, 333–348, 350
Photoshop, 365
Pigott, E., 218
Pillars of creation, 78
Pinwheel, 130, 216, 219, 249, 313, 339
Pisces, 132, 248
Pixel, 356, 357, 365
Pixelated, 365
Pixelation, 365
Planet, 6–8, 11–12, 16, 18, 19, 31, 41, 96, 200, 241
Planetary nebula, 112, 113, 199, 253, 254, 315, 316
Pleiades, 25, 107, 134, 152, 163, 165
Plossl eyepiece, 132, 226
Plough, 38, 150, 165, 338
Pluto, 25, 57
Pollution, 26, 316, 319, 322
Pollux, 160
Porrima, 334
Post processing, 365
Praesepe, 159
Preface, vii–viii
Primary mirror, 358, 365, 366
Procyon, 172, 178
Progenitor star, 112, 253
Ptolemy, 51
Ptolemy's cluster, 51
Pupil, 360
Pupil width, 360
Puppis, 167, 169, 170, 304, 305

R
Radiation, 256
Radio, 256, 272, 275, 338
   waves, 338

Red giant, 112
Reflection nebula, 90
Reflector/refractor
   short focal length, 25, 64, 127, 162
   short tube, 26, 64, 125, 135, 162, 165
Regulus, 160, 314
Resolution, 205, 210, 292, 347
RGB, 39, 132, 222, 270, 273, 327
Rigel, 118
Ring, 2, 112, 113, 199
Ring Nebula, 26, 195, 198, 254
Rosette Nebula, 55
RR Lyrae, 37

S
Sagitta, 113, 239
Sagittarius, 35–36, 48, 49, 52, 55, 92, 94, 95, 98, 101, 103, 105, 106, 108, 115, 117, 189, 191–193, 228, 233–236, 238, 241, 250, 252
Sagittarius Dwarf Elliptical Galaxy, 189
Satellite galaxy, 128, 249, 326, 348
Saturn, 8, 12, 13, 18, 34, 98
SBIG ST2000XCM, 39, 76, 140, 179, 227, 308
Schmidt-Cassegrain, 222, 224
Scorpio, 48, 51, 213, 267
Scorpius, 41, 47, 49, 51, 52, 89, 265, 267, 341
Scutum, 63, 109
Secondary mirror, 12, 29
Semi-apochromatic, 127, 197, 241, 303
Semiconductor, 358
Serpens, 44, 46, 66, 73, 79
Serpens Caput, 44
Serpens Cauda, 78
Serpent, 46, 66, 73
Seven Sisters, 163
Seyfert, C.K., 256
Seyfert Galaxy, 256
Short tube refractor, 26, 64, 125, 135, 162, 165
Sirius, 152, 153, 167, 178
Sleeping Beauty Galaxy, 218
Solar astronomy, 2, 7, 8, 16, 19, 26, 277
Solar system object, 31, 34, 41
Solar system viewing, 25
Sombero Galaxy, 334
Southern Cross, 134
Spherical aberration, 358, 366
Spica, 275, 334
Spindle Galaxy, 23, 24, 330
Spiral arm, 82, 126, 128, 132, 216, 217, 219, 276, 294, 308, 310, 324, 326, 327, 339, 346

Spiral galaxy, 55, 125, 130, 131, 181, 201, 202, 210, 215, 216, 218, 219, 221, 224, 248, 256, 257, 268, 269, 275, 276, 288, 289, 294, 295, 297, 298, 307, 310, 312, 313, 318, 319, 321, 323, 324, 326, 334, 338, 339, 343, 344, 346
Spiral structure, 125, 181, 201, 219, 249, 258, 268–270, 310, 319, 322, 324, 326, 347
Star
   atlas, 23, 209, 248, 301
   cloud, 105
   formation, 78, 268, 294, 307, 323, 338
   variable, 37, 124
Star cluster
   globular, 12, 26, 27, 34, 35, 37–39, 41, 42, 44, 45, 57, 58, 60, 61, 65, 66, 68, 69, 71, 73, 75, 76, 87, 88, 94, 96, 97, 115–117, 121, 122, 125, 128, 130, 187–190, 192, 195, 196, 207, 212–214, 226, 228, 229, 233, 236, 237, 239, 240, 242, 243, 250, 251, 263–267, 285, 301, 302, 334, 337, 341, 349
   open, 14, 26, 27, 47–49, 51–53, 63, 78, 85, 86, 94, 95, 101, 102, 106, 107, 109, 110, 118, 119, 132, 134, 138, 139, 141–146, 148, 152, 153, 160, 163, 164, 167, 168, 170–173, 178, 179, 184, 187, 218, 226, 227, 246, 304, 305, 331, 332
Star-forming region, 53, 82, 130, 155, 221
Stark Software, 76, 135, 140, 158, 163, 186, 217, 227, 279, 288, 324
Starlight Xpress, 39, 120, 132, 142, 156, 182, 222, 270, 273, 327
Star merger, 226, 265, 307
Startravel 80 short tube refractor, 26, 162
Suburban location, 33, 60, 153, 156, 160, 165, 204, 258
Summer Triangle, 239
Sun, 2, 6, 9, 10, 12, 18, 27, 37, 112, 198
Sunflower Galaxy, 2115
Supernova, 10, 31, 268, 277
   remnant, 24, 32, 253
Surface brightness, 49, 224
SXV H9 camera, 80, 156

T
Taurus, 10, 31, 33, 138, 163
Telescope
   achromat, 185, 191, 255
   apochromatic, 127

   Maksutov-Cassegrain, 25, 28, 76, 132, 135, 198, 222, 261, 326
   reflector, 8, 9, 25, 153
   refractor
      achromat, 191
      achromatic, 191
      apochromatic, 127
Terminal light path, 366
Thomas, Oswald, 172
Toucam, 39, 132, 182, 222, 270, 273, 327
Transistor, 6, 9, 12, 19, 29
Trapezium, 156
Triangulum, 130
Triangulum Galaxy, 130
Triffid Nebula, 90, 94
Tripod, 64

U
Unaided eye, 25, 37, 75, 81, 96, 124, 125, 130, 134, 138, 139, 147, 152, 156, 159, 160, 163, 165, 173, 304
Universe, 17, 26, 37, 68
Uranus, 19, 24, 25
Ursa Major, 150, 151, 165, 180, 182, 215, 268–270, 272, 315, 316, 326, 338, 343, 346
Usual Suspects, 139, 142, 144, 152

V
Variable star, 37
Venus, 10, 12, 31
Vesto Slipher, 124
Virgo, 17, 45, 175, 202, 205, 207, 279, 282, 284, 287, 289, 292, 295, 298, 314, 319, 321, 324
Virgo galaxy cluster, 175, 204, 277, 283, 285, 291, 294, 308, 334
Virgo Supercluster, 17, 175, 176, 201, 204, 207, 209, 210, 218, 221, 224, 230, 256, 277, 279, 280, 283, 285, 287, 288, 291, 294, 297, 308, 318, 321, 323, 324, 334
Visibility, 8, 27, 153, 168, 316
Vulpecula, 113, 114, 241

W
Wavelength, 256, 272, 275
Webcam, 367
Web camera, 358, 367
Website, 55
Whirlpool Galaxy, 180, 215
White dwarf, 112, 198, 253

Wide field of view, 91, 103, 127, 153, 156, 169, 171, 221, 224, 278, 284, 326, 333, 339, 343
Wild Duck Cluster, 63

Z

Zenith, 118, 131, 153, 180, 217, 219, 347
Zodiac, 355, 357, 364, 367, 368

Printed in Great Britain
by Amazon